PURE MATHEMATICS

6. COORDINATE GEOMETRY IN TWO DIMENSIONS

Second Edition
By
Anthony Nicolaides

P.A.S.S. PUBLICATIONS
Private Academic & Scientific Studies Limited

© A. NICOLAIDES 1994, 2007

First Published in Great Britain 1991 by
Private Academic & Scientific Studies Limited

ISBN–13   978–1–872684–68–0 £9–95

SECOND EDITION 2008

This book is copyright under the Berne Convention.

All rights are reserved. Apart as permitted under the Copyright Act, 1956, no part of this publication may be reproduced, stored in a retrieval system, or transmitted in any form of by any means, electronic, electrical, mechanical, optical, photocopying, recording or otherwise, without the prior permission of the publisher.

Titles by the same author.

Revised and Enhanced

1. Algebra.   GCE A Level   ISBN–13   978–1–872684–82–6   £11–95

2. Trigonometry.   GCE A Level   ISBN–13   978–1–872684–87–1   £11–95

3. Complex Numbers.   GCE A Level   ISBN–13   978–1–872684–92–5   £9–95

4. Differential Calculus and Applications.   GCE A Level   ISBN–13   978–1–872684–97–0   £9–95

5. Cartesian and Polar Curve Sketching.   GCE A Level   ISBN–13   978–1–872684–63–5   £9–95

6. Coordinate Geometry in two Dimensions.   GCE A Level   ISBN–13   978–1–872684–68–0   £9–95

7. Integral Calculus and Applications.   GCE A Level   ISBN–13   978–1–872684–73–4   £14–95

8. Vectors in two and three dimensions.   GCE A Level   ISBN–13   978–1–872684–15–4   £9–95

9. Determinants and Matrices.   GCE A Level   ISBN–13   978–1–872684–16–1   £9–95

10. Probabilities.   GCE A Level   ISBN–13   978–1–872684–17–8   £8–95
    This book includes the full solutions

11. Success in Pure Mathematics: The complete works of GCE A Level. (1–9 above inclusive)   ISBN   978–1–872684–93–2   £39–95

12. Electrical & Electronic Principles. First year Degree Level   ISBN–13   978–1–872684–98–7   £16–95

13. GCSE Mathematics Higher Tier Third Edition.   ISBN–13   978–1–872684–69–7   £19–95

All the books have answers and a CD is attached with FULL SOLUTIONS of all the exercises set at the end of the book.

Preface

This book, which is part of the GCE A level series in Pure Mathematics covers the specialized topic of Coordinate Geometry in two dimensions.

The GCE A level series success in Pure Mathematics is comprised of nine books, covering the syllabuses of most examining boards. The books are designed to assist the student wishing to master the subject of Pure Mathematics. The series is easy to follow with minimum help. It can be easily adopted by a student who wishes to study it in the comforts of his home at his pace without having to attend classes formally; it is ideal for the working person who wishes to enhance his knowledge and qualification. The Coordinate Geometry book, like all the books in the series, the theory is comprehensively dealt with, together with many worked examples and exercises. A step by step approach is adopted in all the worked examples. A CD is attached to the book with FULL SOLUTIONS of all the exercises set at the end of each chapter.

This book develops the basic concepts and skills that are essential for the GCE A level in Pure Mathematics.

The modules C1, C2, C3, C4, FP1, FP2, FP3 are dealt with adequately in this book.

A. Nicolaides

6. COORDINATE GEOMETRY IN TWO DIMENSIONS

CONTENTS

1. THE STRAIGHT LINE GRAPH — 1

The gradient of a straight line. — 1
The intercept form of a straight line. — 3
The angle between two straight lines. — 5
Intersection between two lines. — 7
Division of a straight line in a given ratio. — 7
Area of a triangle. — 8
Midpoint between two points $A(x_1, y_1)$, $B(x_2, y_2)$. — 8
Distance of a known point from a known straight line. — 10
The perpendicular distance from a point (x_1, y_1) to a straight line $ax + by + c = 0$ (Alternatively). — 10
Summary of formulae of a straight line. — 11
Exercises 1. — 13

CONIC SECTIONS — 17

2. THE CIRCLE — 18

The general equation of a circle. — 18
Condition that a straight line is tangent to a circle. — 19
Orthogonal circles. — 20
Parametric equations of a circle. — 21
Exercises 2. — 23

3. PARABOLAS — 25

Equations of the parabola $y^2 = 4ax$. — 25
$y^2 = -4ax$. — 25
$x^2 = 4ay$. — 25
$x^2 = -4ay$. — 25
Definition of a parabola. Parameters. — 26
Gradient of chord PQ. — 26
Equation of tangent to $x^2 = 4ay$. — 26
Equation of tangent at $P(x_1, y_1)$ on $x^2 = 4ay$. — 27
Equation of normal at $P(2ap, ap^2)$. — 27
Point of intersection of the tangents at $P(2ap, ap^2)$ and $Q(2aq, aq^2)$. — 27
Point of intersection of the normals at $P(2ap, ap^2)$ and $Q(2aq, aq^2)$. — 28

	contents
The focal chord.	28
Condition for a straight line $y = mx + c$ to touch $x^2 = 4ay$.	28
Equation of the chord of contact from (x_1, y_1) to $x^2 = 4ay$.	29
Properties of the parabola.	30
Exercises 3.	31

4. THE ELLIPSE — 33

To derive the equation of the ellipse.	33
Eccentric angles.	34
Determine the equation of the tangent at $P(x_1, y_1)$ to the ellipse $\frac{x^2}{a^2} + \frac{y^2}{b^2} = 1$.	35
Determine the tangent to the ellipse $\frac{x^2}{a^2} + \frac{y^2}{b^2} = 1$ at the point $P(a\cos\theta, b\sin\theta)$.	35
Determine the normal to the ellipse.	36
Determine the equation of the chord of the ellipse $\frac{x^2}{a^2} + \frac{y^2}{b^2}$ between the points $P(a\cos\theta, b\sin\theta)$ and $Q(a\cos\phi, b\sin\phi)$.	36
Exercises 4.	37

5. THE HYPERBOLA — 39

Definition Derivation of the equation of the hyperbola.	39
The parametric equations of the hyperbola Trigonometric.	39
Hyperbolic.	40
The equation of the tangent at the point $P(a\sec\theta, b\tan\theta)$.	40
The equation of the normal at the point $P(a\sec\theta, b\tan\theta)$.	40
The equation of the tangent at the point $P(x_1, y_1)$ of the hyperbola.	40
Asymptotes to the hyperbola $\frac{x^2}{a^2} - \frac{y^2}{b^2} = 1$.	41
The rectangular hyperbola $x^2 - y^2 = a^2$.	44
The rectangular hyperbola $xy = c^2$.	46
Conjugate axis.	46
The equation of the tangent to the hyperbola $xy = c^2$ at the point $P(x_1, y_1)$.	46
The parametric equations of the rectangular hyperbola $xy = c^2$ Equation of the tangent at $P\left(ct, \frac{c}{t}\right)$.	47
The equation of the normal to the hyperbola $xy = c^2$ to the point $P\left(ct, \frac{c}{t}\right)$.	48
The equation of the chord joining the points $P\left(cp, \frac{c}{p}\right)$ and $Q\left(cq, \frac{c}{q}\right)$.	49
The locus of the point $x = ct$, $y = \frac{c}{t}$.	50
The properties of the rectangular hyperbola $xy = c^2$.	50
The area between the tangent and the axes.	50

— **GCE A level**

The equation of a rectangular hyperbola referred to axes parallel to
the asymptotes through the point (h, k). 51

The equation of chord through $P_1\left(ct_1, \frac{c}{t_1}\right)$ and $P_2\left(ct_2, \frac{c}{t_2}\right)$. 51

The diameter of the rectangular hyperbola $xy = c^2$. 52

Conjugate hyperbolas. 53

Central rectangle of the hyperbola. 56

Exercises 5. 58

MISCELLANEOUS 60

ANSWERS 68

INDEX 72

The Straight Line Graph

The general form of a straight line is given by $ax + by + c = 0$ where a, b and c are constants and x, y are the variables

$$ax + by + c = 0$$

The Gradient of a Straight Line

A line passes through two fixed points (x_1, y_1) and (x_2, y_2) as shown in Fig. 6-I/1

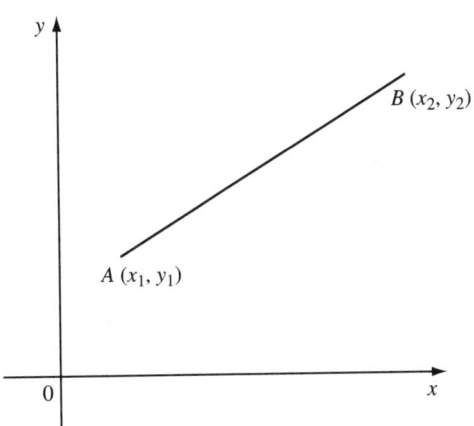

Fig. 6-I/1

The length $AB = \sqrt{(y_2 - y_1)^2 + (x_2 - x_1)^2}$ using Pythagoras. The gradient of the straight line is denoted by m and it is given as

$$m = \frac{y_2 - y_1}{x_2 - x_1}$$

(x_1, y_1) and (x_2, y_2) are two sets of coordinates x_1, x_2 are the horizontal distances from a fixed point, the origin, the intersection of two perpendicular axes, the x-axis and y-axis of **DESCARTES**, the cartesian axes, and y_1, y_2 are the vertical distances from the fixed point, O.

x_1 and x_2 are called <u>abscissae</u> and y_1 and y_2 are called <u>ordinates</u>.

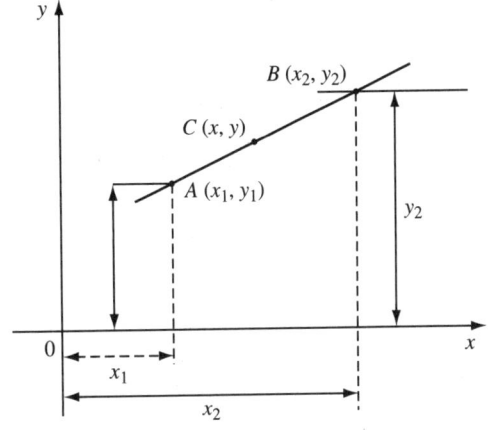

Fig. 6-I/2

The coordinates of the origin are given as $O(0, 0)$. If (x, y) is any point on the straight line, the gradient will be $\frac{y - y_1}{x - x_1} = m$ and the equation of the straight line is

$$\frac{y - y_1}{x - x_1} = \frac{y_2 - y_1}{x_2 - x_1} = m$$

WORKED EXAMPLE 1

A straight line passes through the points $(2, 3)$ and $(-1, -4)$, determine the equation of the line.

2 — GCE A level

Solution 1

(a) $m = \dfrac{-4 - 3}{-1 - 2} = \dfrac{-7}{-3} = \dfrac{7}{3}$.

If $y = mx + c$ is the gradient/intercept form, the line is then $y = \dfrac{7}{3}x + c$, it passes through (2, 3),

$3 = \dfrac{7}{3} \cdot 2 + c$

$c = 3 - \dfrac{14}{3} = -\dfrac{5}{3}$

$y = \dfrac{7}{3}x - \dfrac{5}{3}$

$\boxed{3y - 7x + 5 = 0}$ ✓

(b) Alternatively, by using $\dfrac{y - y_1}{x - x_1} = \dfrac{y_2 - y_1}{x_2 - x_1}$

$\dfrac{y - 3}{x - 2} = \dfrac{-4 - 3}{-1 - 2}$

$\dfrac{y - 3}{x - 2} = \dfrac{7}{3} \Rightarrow 3y - 9 = 7x - 14$

$\boxed{3y - 7x + 5 = 0}$ ✓

The gradient/intercept form of a straight line is given by

$\boxed{y = mx + c}$

where m is the gradient of the line and c is the y-intercept of the line. The gradient-intercept form of a straight line is the most useful form since at a glance we can locate the line.

WORKED EXAMPLE 2

Sketch the lines

(i) $y = x + 1$

(ii) $y = -x + 2$

(iii) $y = 2x - 3$.

Solution 2

(i) $y = x + 1$ when $x = 0$, $y = 1$, the line passes through the point (0, 1) and has a gradient of $m = 1$.

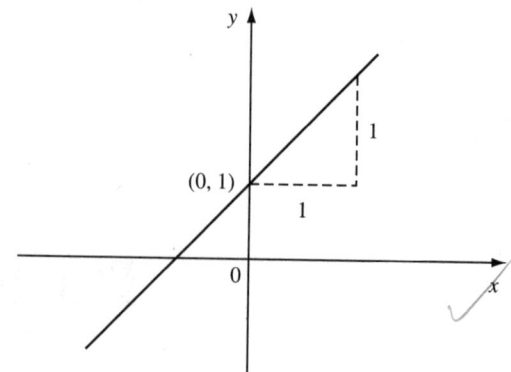

Fig. 6-I/3

(ii) $y = -x + 2$ when $x = 0$, $y = 2$, the line passes through the point (0, 2) and has a gradient of $m = -1$.

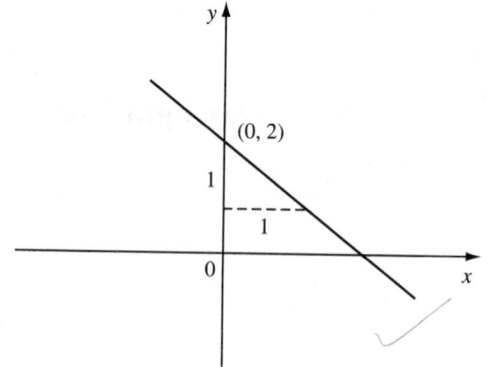

Fig. 6-I/4

(iii) $y = 2x - 3$ when $x = 0$, $y = -3$, the line passes through the point, (0, −3) and has a gradient of $m = 2$.

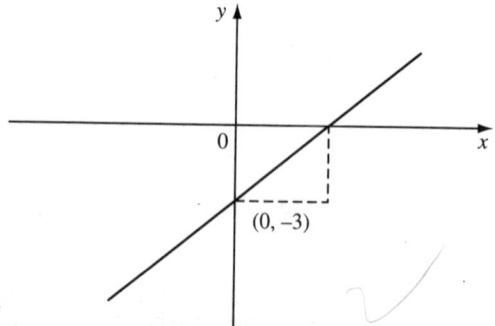

Fig. 6-I/5

If $m = 0$ the gradient/intercept form of the straight line becomes $\boxed{y = c}$

WORKED EXAMPLE 3

Sketch the lines (i) $y = 2$ (ii) $y = 0$ (iii) $y = -3$.

Solution 3

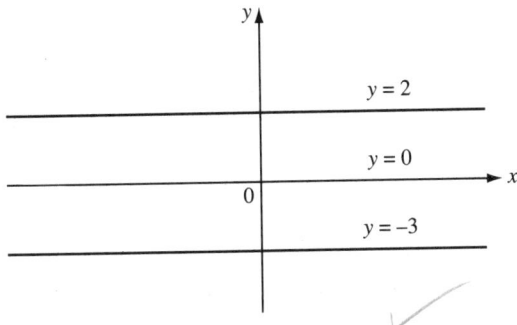

Fig. 6-I/6

The straight lines are parallel to the x-axis. If $m = \infty$, the gradient/intercept form of the straight line becomes $y = mx + c$, $\frac{y}{m} = x + \frac{c}{m}$ $x = 0$ or $x = k$ where x is a constant.

WORKED EXAMPLE 4

Sketch the lines (i) $x = -2$ (ii) $x = 0$ and (iii) $x = 3$.

Solution 4

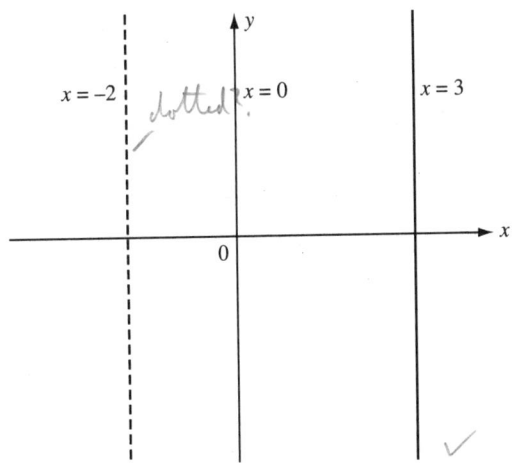

Fig. 6-I/7

The straight lines are parallel to the y-axis. If the intercept, c, is zero the straight lines pass through the origin.
$$y = mx$$

WORKED EXAMPLE 5

Sketch the line (i) $y = x$

(ii) $y = -x$

(iii) $y = 2x$

(iv) $y = -3x$.

Solution 5

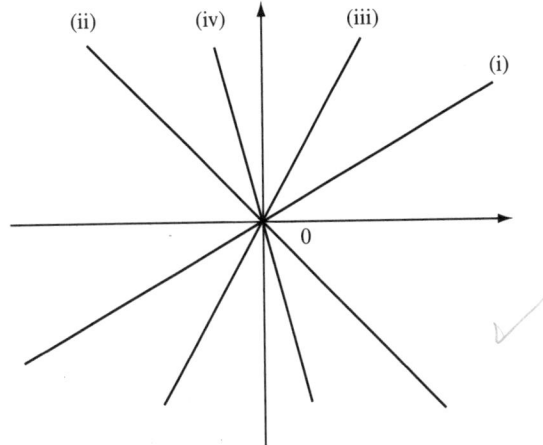

Fig. 6-I/8

The Intercept Form of a Straight Line

The intercept form of a straight line is given as:

$$\boxed{\frac{x}{a} + \frac{y}{b} = 1}$$

If $y = 0, x = a$; if $x = 0, y = b$

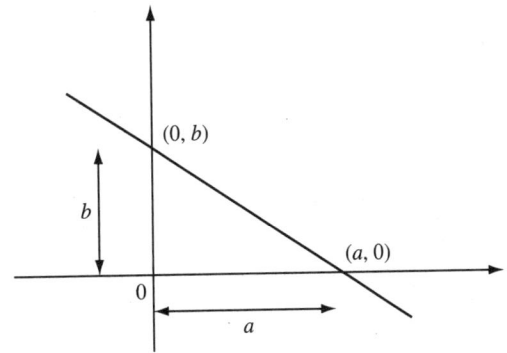

Fig. 6-I/9

4 — GCE A level

WORKED EXAMPLE 6

Sketch the straight lines:

(i) $\dfrac{x}{1} + \dfrac{y}{2} = 1$

(ii) $\dfrac{x}{-2} + y = 1$

(iii) $\dfrac{x}{3} + \dfrac{y}{-2} = 1$.

Solution 6

(i) $\dfrac{x}{1} + \dfrac{y}{2} = 1$

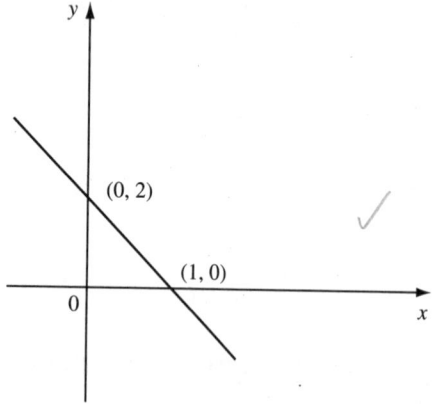

Fig. 6-I/10

(ii) $\dfrac{x}{-2} + y = 1$

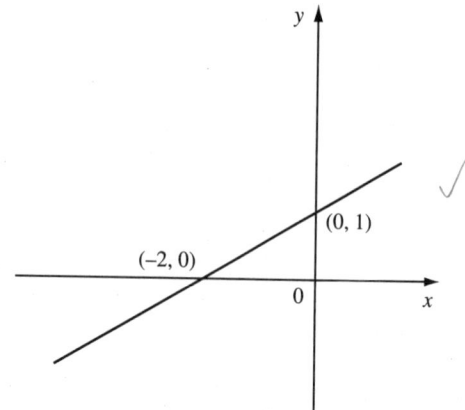

Fig. 6-I/11

(iii) $\dfrac{x}{3} + \dfrac{y}{-2} = 1$

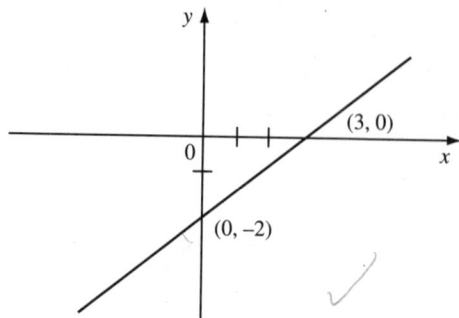

Fig. 6-I/12

WORKED EXAMPLE 7

The following straight lines have the general form

(i) $3x + 2y - 1 = 0$

(ii) $x - y + 4 = 0$

(iii) $-2x + 5y - 4 = 0$.

Give (a) the gradient/intercept form of the lines
and (b) the intercept form of the lines.

Solution 7

(a) (i) $3x + 2y - 1 = 0$

$2y = -3x + 1$ $\boxed{y = -\dfrac{3}{2}x + \dfrac{1}{2}}$

(ii) $x - y + 4 = 0$ $\boxed{y = x + 4}$

(iii) $-2x + 5y - 4 = 0$

$5y = 2x + 4$

$y = \dfrac{2}{5}x + \dfrac{4}{5}$

(b) (i) $3x + 2y - 1 = 0$

$3x + 2y = 1$ $\boxed{\dfrac{x}{\frac{1}{3}} + \dfrac{y}{\frac{1}{2}} = 1}$

(ii) $x - y + 4 = 0$

$x - y = -4$ $\boxed{\dfrac{x}{-4} + \dfrac{y}{4} = 1}$

Make x = 0, y = 0

(iii) $-2x + 5y - 4 = 0$

$-2x + 5y = 4$

$\boxed{\dfrac{x}{-2} + \dfrac{y}{\frac{4}{5}} = 1}$ ✓

The Angle between Two Straight Lines

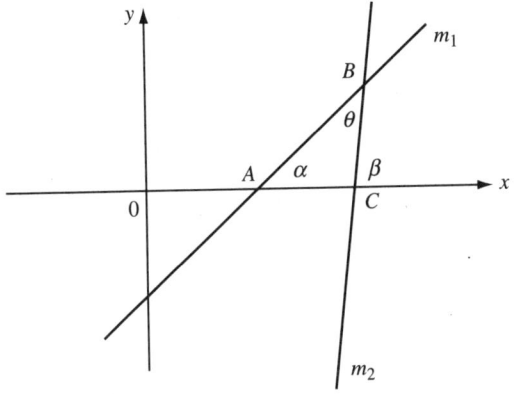

Fig. 6-I/13

Let θ be the acute angle between two lines of gradients m_1 and m_2 as shown in the diagram of Fig. 6-I/13.

If α is the angle that m_1 line makes with the horizontal and if β is the angle that m_2 line makes with the horizontal then

$\tan \alpha = m_1$ and $\tan \beta = m_2$.

From the triangle ABC, we have $\theta + \alpha + 180 - \beta = 180$

$\boxed{\theta = \beta - \alpha}$

taking tangents on both sides, we have

$\tan \theta = \tan(\beta - \alpha)$

$= \dfrac{\tan \beta - \tan \alpha}{1 + \tan \beta \tan \alpha}$

$\boxed{\tan \theta = \dfrac{m_2 - m_1}{1 + m_1 m_2}}$

If $\theta = 0$, the two straight lines are parallel and $m_2 - m_1 = 0$,

$\boxed{m_2 = m_1}$

Therefore, two straight lines are parallel if the gradients are equal.

If $\theta = 90°$, the two straight lines are perpendicular, $\dfrac{m_2 - m_1}{1 + m_1 m_2} = \infty$ that is, $1 + m_1 m_2 = 0$ or

$\boxed{m_1 m_2 = -1}$

The angle between the two lines is given

$\boxed{\theta = \tan^{-1}\left(\dfrac{m_2 - m_1}{1 + m_1 m_2}\right)}$

If θ is acute m_2 is greater than m_1 if θ is obtuse m_1 is greater than m_2 provided $1 + m_1 m_2$ is positive.

WORKED EXAMPLE 8

Find the angle between the lines

(i) $3x - 5y = 2$ and $x + 4y = 5$

(ii) $x + y = 1$ and $2x - 3y = 4$.

Solution 8

(i) $3x - 5y = 2$, $5y = 3x - 2$, $y = \dfrac{3}{5}x - \dfrac{2}{5}$

$x + 4y = 5$, $4y = -x + 5$, $y = -\dfrac{1}{4}x + \dfrac{5}{4}$.

The gradients of the lines are $\dfrac{3}{5}$ and $-\dfrac{1}{4}$ respectively, if $m_1 = -\dfrac{1}{4}$ and $m_2 = \dfrac{3}{5}$ then

$\theta = \tan^{-1}\left(\dfrac{m_2 - m_1}{1 + m_1 m_2}\right)$

$= \tan^{-1} \dfrac{\dfrac{3}{5} - \left(-\dfrac{1}{4}\right)}{1 + \left(-\dfrac{1}{4}\right)\left(\dfrac{3}{5}\right)}$

$= \tan^{-1} \dfrac{\frac{17}{20}}{\frac{17}{20}}$

$= \tan^{-1} 1 = 45°$

if $m_1 = \dfrac{3}{5}$, $m_2 = -\dfrac{1}{4}$,

$\theta = \tan^{-1} \dfrac{\left(-\dfrac{1}{4}\right) - \left(\dfrac{3}{5}\right)}{1 - \dfrac{3}{20}}$

$= \tan^{-1} -\dfrac{\frac{17}{20}}{\frac{17}{20}} = 135°$

6 — GCE A level

(ii) $x + y = 1$ and $2x - 3y = 4$

$y = -x + 1$ $\qquad 3y = 2x - 4$

or $y = \dfrac{2}{3}x - \dfrac{4}{3}$.

The gradients of the lines are -1 and $\frac{2}{3}$ respectively, if $m_2 = \frac{2}{3}$ and $m_1 = -1$.

$$\theta = \tan^{-1}\dfrac{m_2 - m_1}{1 + m_1 m_2} = \tan^{-1}\dfrac{\frac{2}{3} - (-1)}{1 - \frac{2}{3}}$$

$$= \tan^{-1}\dfrac{\frac{5}{3}}{\frac{1}{3}} = \tan^{-1} 5$$

$= 78.69°$ is the acute angle

if $m_1 = \dfrac{2}{3}, m_2 = -1$

$$\theta = \tan^{-1}\dfrac{-1 - \frac{2}{3}}{1 - \frac{2}{3}} = \tan^{-1}\dfrac{-\frac{5}{3}}{\frac{1}{3}}$$

$$= \tan^{-1} -5$$

$= 101.31°$ is the obtuse angle.

Any two lines have an acute angle and an obtuse angle between them

$\theta =$ acute

$180° - \theta =$ obtuse

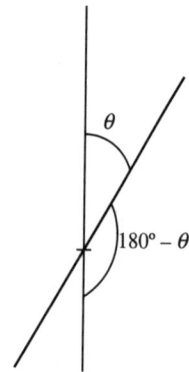

Fig. 6-I/14

unless the angle between them is 90°.

Worked Example 9

Find the tangent of each angle of the triangle having the given vertices $A(2, 3), B(-2, 5), C(3, -4)$.

Solution 9

The gradient of $AB = \dfrac{5 - 3}{-2 - 2}$

$$m_1 = -\dfrac{1}{2}.$$

The gradient of $AC = \dfrac{3 - (-4)}{2 - 3}$

$$m_2 = \dfrac{7}{-1} = -7.$$

The gradient of $BC = \dfrac{5 - (-4)}{-2 - 3}$

$$m_3 = -\dfrac{9}{5}.$$

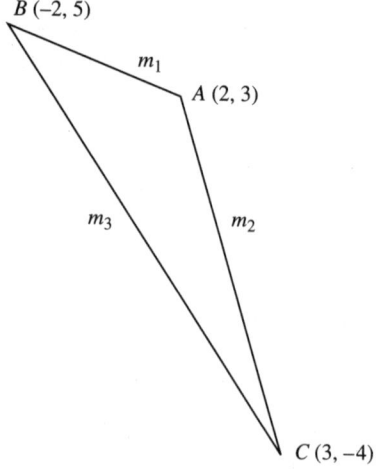

Fig. 6-I/15

The tangent of the angle between AB and AC

$$\tan A = \dfrac{-7 - \left(-\frac{1}{2}\right)}{1 + (-7)\left(-\frac{1}{2}\right)} = \dfrac{-\frac{13}{2}}{\frac{9}{2}} = -\dfrac{13}{9}$$ the angle is obtuse.

The tangent of the angle between AB and BC

$$\tan B = \frac{-\frac{1}{2}-\left(-\frac{9}{5}\right)}{1+\left(-\frac{1}{2}\right)\left(-\frac{9}{5}\right)} = \frac{\frac{9}{5}-\frac{1}{2}}{1+\frac{9}{10}}$$

$$= \frac{\frac{13}{10}}{\frac{19}{10}} = \frac{13}{19}.$$

The tangent of the angle between BC and AC

$$\tan C = \frac{-\frac{9}{5}-(-7)}{1+\left(-\frac{9}{5}\right)(-7)} = \frac{7-\frac{9}{5}}{1+\frac{63}{5}} = \frac{\frac{26}{5}}{\frac{68}{5}} = \frac{13}{34}.$$

Intersection Between Two Lines

Worked Example 10

To find the point of intersection between the lines

$$x + y = 3 \qquad \ldots (1)$$
$$-x + 2y = 0 \qquad \ldots (2)$$

Solution 10

Solving the simultaneous equations. Adding (1) and (2)

$$3y = 3 \quad y = 1$$
$$x + 1 = 3, \quad x = 2.$$

The lines intersect at $P(2, 1)$

To find the coordinates of the point of intersection, we solve the two linear equations simultaneously.

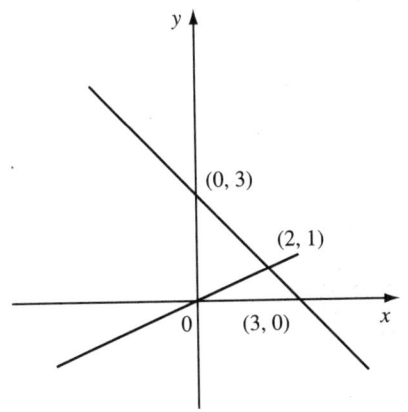

Fig. 6-I/16

Division of a Straight Line in a Given Ratio

Let $A(x_1, y_1)$ and $B(x_2, y_2)$ be two points on a straight line.

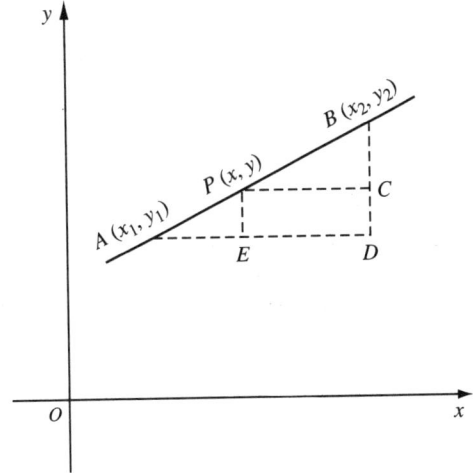

Fig. 6-I/17

Let $P(x, y)$ be a point that divides the line AB in the ratio $\lambda : \mu$.

The triangles APE and PBC are similar, therefore

$$\frac{AP}{PB} = \frac{AE}{PC} = \frac{PE}{BC} = \frac{\lambda}{\mu}$$

$$\frac{AE}{PC} = \frac{x - x_1}{x_2 - x} = \frac{\lambda}{\mu} \text{ therefore } \mu(x - x_1) = \lambda(x_2 - x)$$

$$x\mu + \lambda x = \lambda x_2 + \mu x_1$$

$$\boxed{x = \frac{\lambda x_2 + \mu x_1}{\lambda + \mu}} \qquad \ldots (1)$$

$$\frac{PE}{BC} = \frac{y - y_1}{y_2 - y} = \frac{\lambda}{\mu} \text{ therefore } \mu(y - y_1) = \lambda(y_2 - y)$$

$$\mu y + \lambda y = \lambda y_2 + \mu y_1$$

$$\boxed{y = \frac{\lambda y_2 + \mu y_1}{\lambda + \mu}} \qquad \ldots (2)$$

Equation (1) and (2) can be quoted. The formulae are also applicable if the line is divided in the same ratio $\lambda : \mu$ externally.

Area of Triangle

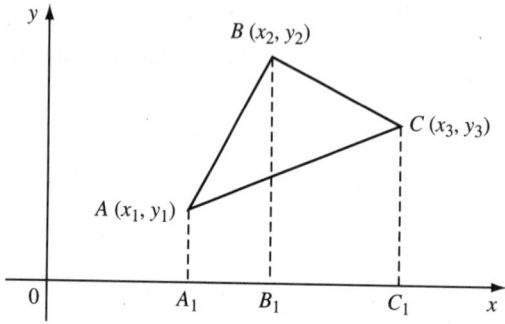

Fig. 6-I/18

Area of triangle ABC = area of trapezium A_1ABB_1 + area of trapezium B_1BCC_1 − area of the trapezium A_1ACC_1.

Area of trapezium $A_1ABB_1 = \dfrac{1}{2}(A_1A + BB_1)A_1B_1$

$$= \dfrac{1}{2}(y_1 + y_2)(x_2 - x_1)$$

Area of trapezium $B_1BCC_1 = \dfrac{1}{2}(BB_1 + CC_1)B_1C_1$

$$= \dfrac{1}{2}(y_2 + y_3)(x_3 - x_2)$$

Area of trapezium $A_1ACC_1 = \dfrac{1}{2}(A_1A + CC_1)A_1C_1$

$$= \dfrac{1}{2}(y_1 + y_3)(x_3 - x_1).$$

Area of triangle ABC

$$= \dfrac{1}{2}(y_1 + y_2)(x_2 - x_1) + \dfrac{1}{2}(y_2 + y_3)(x_3 - x_2)$$

$$- \dfrac{1}{2}(y_1 + y_3)(x_3 - x_1)$$

$$= \dfrac{1}{2}y_1x_2 + \dfrac{1}{2}y_2x_2 - \dfrac{1}{2}x_1y_1 - \dfrac{1}{2}x_1y_2$$

$$+ \dfrac{1}{2}y_2x_3 + \dfrac{1}{2}x_3y_3 - \dfrac{1}{2}x_2y_2$$

$$- \dfrac{1}{2}x_2y_3 - \dfrac{1}{2}y_1x_3 - \dfrac{1}{2}y_3x_3 + \dfrac{1}{2}x_1y_1 + \dfrac{1}{2}x_1y_3.$$

Area of $\triangle ABC = \left| \dfrac{1}{2}y_1x_2 - \dfrac{1}{2}x_1y_2 + \dfrac{1}{2}x_1y_3 + \dfrac{1}{2}y_2x_3 \right.$

$$\left. - \dfrac{1}{2}x_2y_3 - \dfrac{1}{2}y_1x_3 \right|$$

Area of $\triangle ABC$

$$= \dfrac{1}{2}(x_1y_2 - y_1x_2 + x_2y_3 - x_3y_2 + x_3y_1 - x_1y_3).$$

This is very difficult to memorize

$$\text{Area } \triangle ABC = \dfrac{1}{2}\begin{vmatrix} 1 & 1 & 1 \\ x_1 & x_2 & x_3 \\ y_1 & y_2 & y_3 \end{vmatrix}$$

$$= \dfrac{1}{2}\begin{vmatrix} x_2 & x_3 \\ y_2 & y_3 \end{vmatrix} - \dfrac{1}{2}\begin{vmatrix} x_1 & x_3 \\ y_1 & y_3 \end{vmatrix}$$

$$+ \dfrac{1}{2}\begin{vmatrix} x_1 & x_2 \\ y_1 & y_2 \end{vmatrix}$$

$$= \dfrac{1}{2}(x_2y_3 - y_2x_3) - \dfrac{1}{2}(x_1y_3 - x_3y_1)$$

$$+ \dfrac{1}{2}(x_1y_2 - x_2y_1)$$

$$= \dfrac{1}{2}\Big(x_1y_2 - y_1x_2 + x_2y_3 - x_3y_2$$

$$+ x_3y_1 - x_1y_3\Big).$$

Midpoint Between Two Points $A(x_1, y_1)$ and $B(x_2, y_2)$

If $P(x, y)$ is the mid-point between the two points $A(x_1, y_1)$ and $B(x_2, y_2)$, then $\dfrac{\lambda}{\mu} = \dfrac{1}{1}$ and $\lambda = \mu$.

Let $M(x, y)$ be the mid-point, the coordinates are given by putting $\lambda = \mu = 1$ in (1) & (2).

$$\boxed{x = \dfrac{x_1 + x_2}{2}}$$

$$\boxed{y = \dfrac{y_1 + y_2}{2}}$$

WORKED EXAMPLE 11

A triangle ABC has the following coordinates $A(6, -1)$, $B(2, 5)$, $C(-3, 3)$. Determine the coordinates of the mid-points P, Q, R of the three sides AB, BC, AC respectively. Hence determine the area of the triangle PQR.

Solution 11

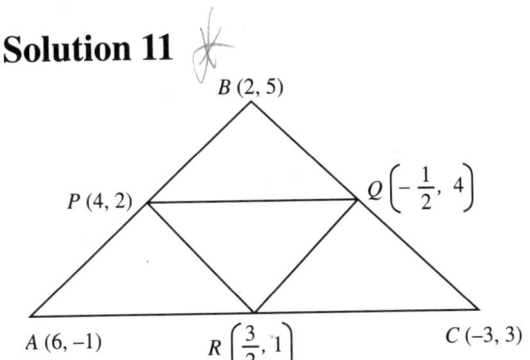

Fig. 6-I/19

$P(x_1, y_1), Q(x_2, y_2), R(x_3, y_3)$

$x_1 = \dfrac{6+2}{2} = 4, \quad y_1 = \dfrac{-1+5}{2} = 2,$

therefore $P(4, 2)$

$x_2 = \dfrac{2+(-3)}{2} = -\dfrac{1}{2}, \quad y_2 = \dfrac{5+3}{2} = 4,$

therefore $Q\left(-\dfrac{1}{2}, 4\right)$

$x_3 = \dfrac{6-3}{2} = \dfrac{3}{2}, \quad y_3 = \dfrac{-1+3}{2} = 1,$

therefore $R\left(\dfrac{3}{2}, 1\right).$

Area $= \sqrt{s(s-a)(s-b)(s-c)}$ Heron's formula where s is the semi-perimeter.

$PQ = \sqrt{\left(4+\dfrac{1}{2}\right)^2 + (2-4)^2} = \sqrt{4.5^2 + 4} = 4.92$

$QR = \sqrt{\left(-\dfrac{1}{2}-\dfrac{3}{2}\right)^2 + (4-1)^2} = \sqrt{4+9} = 3.61$

$PR = \sqrt{\left(4-\dfrac{3}{2}\right)^2 + (2-1)^2} = \sqrt{6.25+1} = 2.69$

$s = \dfrac{1}{2}\left(4.92 + 3.61 + 2.69\right) = 5.61.$

Area $= \sqrt{5.61(5.61-4.92)(5.61-3.61)(5.61-2.69)}$

$= \sqrt{5.61 \times 0.69 \times 2 \times 2.92}$

$= 4.75$ square units.

Area $= \dfrac{1}{2}\begin{vmatrix} 1 & 1 & 1 \\ x_1 & x_2 & x_3 \\ y_1 & y_2 & y_3 \end{vmatrix}$

$= \dfrac{1}{2}\left(\begin{vmatrix} x_2 & x_3 \\ y_2 & y_3 \end{vmatrix} - \begin{vmatrix} x_1 & x_3 \\ y_1 & y_3 \end{vmatrix} + \begin{vmatrix} x_1 & x_2 \\ y_1 & y_2 \end{vmatrix}\right)$

$\dfrac{1}{2}\begin{vmatrix} 1 & 1 & 1 \\ 4 & -\dfrac{1}{2} & \dfrac{3}{2} \\ 2 & 4 & 1 \end{vmatrix}$

$= \dfrac{1}{2}\left(\begin{vmatrix} -\dfrac{1}{2} & \dfrac{3}{2} \\ 4 & 1 \end{vmatrix} - \begin{vmatrix} 4 & \dfrac{3}{2} \\ 2 & 1 \end{vmatrix} + \begin{vmatrix} 4 & -\dfrac{1}{2} \\ 2 & 4 \end{vmatrix}\right)$

$= \dfrac{1}{2}\left[\left(-\dfrac{1}{2} - 6\right) - (4 - 3) + (16 + 1)\right]$

$= 4.75$ square units.

The point of intersection of the medians of a Δ is called the centroid and divides each median in the ratio 2:1.

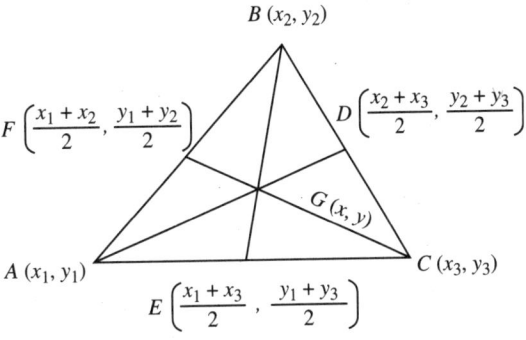

Fig. 6-I/20

$\dfrac{AG}{GD} = \dfrac{BG}{GE} = \dfrac{CG}{GF} = \dfrac{2}{1}$

$\dfrac{AG}{GD} = \dfrac{2}{1} \quad x = \dfrac{\left[2\left(\dfrac{x_2+x_3}{2}\right) + x_1\right]}{3}$

$y = \dfrac{\left[2\left(\dfrac{y_2+y_3}{2}\right) + y_1\right]}{3}$

$G = \left(\dfrac{1}{3}(x_1+x_2+x_3), \dfrac{1}{3}(y_1+y_2+y_3)\right)$

The Distance of a Known Point From a Known Straight Line

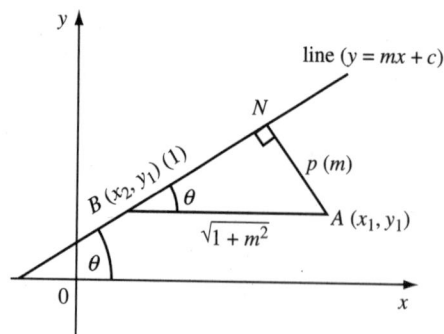

Fig. 6-I/21 $AN = p$

$\tan \theta = \dfrac{m}{1}$ = gradient

$\sin \theta = \dfrac{m}{\sqrt{1+m^2}}$

$AN = p = AB \sin \theta$

$y = mx + c$

$mx = y - c$

$x = \dfrac{y-c}{m} = \dfrac{y_1 - c}{m} = x_2$

$AB = x_1 - x_2 = x_1 - \dfrac{y_1 - c}{m}$

$p = AN = AB \sin \theta$

$= \left(x_1 - \dfrac{y_1 - c}{m} \right) \dfrac{m}{\sqrt{1+m^2}}$

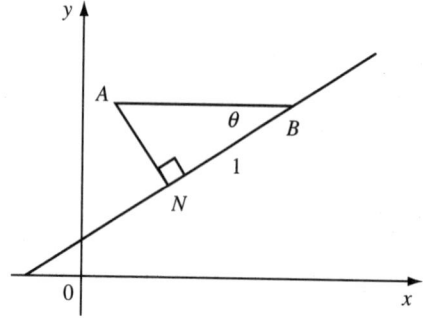

Fig. 6-I/22 $AN = p$

$AN = p = AB \sin \theta = AB \cdot \dfrac{m}{\sqrt{1+m^2}}$

$AB = x_2 - x_1 = \dfrac{y_1 - c}{m} - x_1$

$y = mx + c$

$y_1 = mx_2 + c \quad x_2 = \dfrac{y_1 - c}{m}$

$p = \left(\dfrac{y_1 - c}{m} - x_1 \right) \dfrac{m}{\sqrt{1+m^2}}$

$p = \dfrac{y_1 - c - x_1 m}{\sqrt{1+m^2}}$

$\therefore p = \pm \dfrac{mx_1 - y_1 + c}{\sqrt{1+m^2}}$

Therefore $p = \pm \dfrac{ax_1 + by_1 + c}{\sqrt{a^2 + b^2}}$

Alternatively the Perpendicular Distance from a Point $A(x_1, y_1)$ to a Straight Line $ax + by + c = 0$

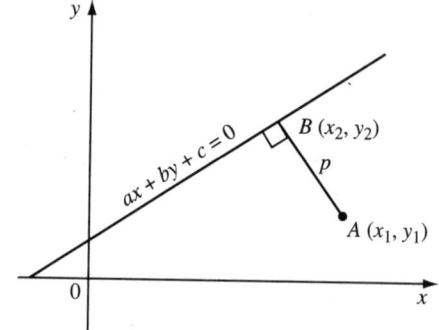

Fig. 6-I/23 $B(x_2, y_2)$

The gradient of the straight line $ax + by + c = 0$ is $m_1 = -\dfrac{a}{b}$, the gradient of AB, is $m_2 = \dfrac{b}{a}$ since $m_1 m_2 = -1$.

The equation of AB, $y = \dfrac{b}{a} x + c$, $y_1 = \dfrac{b}{a} x_1 + c$,

$c = y_1 - \dfrac{b}{a} x_1$,

$y = \dfrac{b}{a} x + y_1 - \dfrac{b}{a} x_1$

The lines $ax + by + c = 0$ and $y = \dfrac{b}{a} x + y_1 - \dfrac{b}{a} x_1$ intersect at $B(x_2, y_2)$.

$$y_2 = \frac{b}{a}x_2 + y_1 - \frac{b}{a}x_1 = -\frac{a}{b}x_2 - \frac{c}{b}$$

$$\frac{b}{a}x_2 + \frac{a}{b}x_2 = -y_1 + \frac{b}{a}x_1 - \frac{c}{b}$$

$$x_2\left(\frac{b^2+a^2}{ab}\right) = \frac{-aby_1 + b^2x_1 - ac}{ab}$$

$$x_2 = \frac{b^2x_1 - aby_1 - ac}{a^2+b^2}$$

$$y_2 = \frac{b}{a}\cdot\frac{(b^2x_1 - aby_1 - ac)}{(a^2+b^2)} + y_1 - \frac{b}{a}x_1$$

$$p = \sqrt{(y_2-y_1)^2 + (x_2-x_1)^2}$$

$$= \sqrt{\left(\dfrac{\dfrac{b^3x_1}{a} - \dfrac{ab^2y_1}{a} - \dfrac{abc}{a} + a^2y_1}{a^2+b^2} +b^2y_1 - \dfrac{a^2bx_1}{a} - \dfrac{b^2bx_1}{a}\right) - y_1\right)^2 + \left(\dfrac{b^2x_1-aby_1-ac}{a^2+b^2} - x_1\right)^2}$$

$$p^2 = \frac{(-bc + a^2y_1 - abx_1 - a^2y_1 - b^2y_1)^2}{(a^2+b^2)^2}$$

$$+ \frac{(b^2x_1 - a^2x_1 - b^2x_1 - aby_1 - ac)^2}{(a^2+b^2)^2}$$

$$= \frac{(bc + abx_1 + y_1b^2)^2 + (a^2x_1 + aby_1 + ac)^2}{(a^2+b^2)^2}$$

$$= \frac{b^2(c + ax_1 + by_1)^2 + a^2(ax_1 + by_1 + c)^2}{(a^2+b^2)^2}$$

$$= \frac{(ax_1 + by_1 + c)^2(a^2+b^2)}{(a^2+b^2)^2}$$

$$\boxed{p = \pm\frac{ax_1 + by_1 + c}{\sqrt{a^2+b^2}}}$$

The Straight Line Graph — 11

Summary

1. The distance between two points $P(x_1, y_1)$ and $Q(x_2, y_2)$

$$\boxed{PQ = \sqrt{(x_2-x_1)^2 + (y_2-y_1)^2}}$$

2. The gradient of the line PQ

$$\boxed{m = \frac{y_2-y_1}{x_2-x_1}}$$

3. The equation of the straight line with gradient m and intercept c

$$\boxed{y = mx + c}$$

4. The general equation of a straight line (a and b not both zero).

$$\boxed{ax + by + c = 0}$$

5. The intercept form of a straight line ($a, b \neq 0$)

$$\boxed{\frac{x}{a} + \frac{y}{b} = 1}$$

6. The equation of a straight line with gradient m and passing through a point $P(x_1, y_1)$

$$\boxed{\frac{y-y_1}{x-x_1} = m}$$

7. The equation of a straight line passing through two points $P(x_1, y_1)$ and $Q(x_2, y_2)$

$$\boxed{\frac{y_2-y_1}{x_2-x_1} = \frac{y-y_1}{x-x_1}}$$

8. The equation of a straight line with x intercept a and parallel to the y-axis

$$\boxed{x = a}$$

9. The equation of a straight line with y intercept b and parallel to the x-axis.

$$\boxed{y = b}$$

10. The perpendicular distance from the point $P(x_1, y_1)$ to the straight line $ax + by + c = 0$

$$\boxed{p = \frac{|ax_1 + by_1 + c|}{\sqrt{a^2+b^2}}}$$

12 — GCE A level

11. Two straight lines with gradients m_1 and m_2 are parallel if

$$\boxed{m_1 = m_2}$$

12. Two straight lines with gradients m_1 and m_2 are perpendicular if

$$\boxed{m_1 m_2 = -1}$$

13. The angle between two lines of gradients m_1 and m_2 ($m_1 m_2 \neq -1$)

$$\boxed{\theta = \tan^{-1} \frac{m_2 - m_1}{1 + m_1 m_2}}$$

14. The coordinates of the mid point between two points $P(x_1, y_1)$ and $Q(x_2, y_2)$

$$\boxed{\left(\frac{x_1 + x_2}{2}, \frac{y_1 + y_2}{2}\right)}$$

15. The meet of the medians (or the centre of gravity) of a triangle ABC with coordinates $A(x_1, y_1)$, $B(x_2, y_2)$, $C(x_3, y_3)$.

$$\boxed{\left(\frac{x_1 + x_2 + x_3}{3}, \frac{y_1 + y_2 + y_3}{3}\right)}$$

16. The area of a triangle ABC with coordinates $A(x_1, y_1)$, $B(x_2, y_2)$ and $C(x_3, y_3)$.

$$\boxed{\text{AREA } \triangle ABC = \frac{1}{2} \begin{vmatrix} 1 & 1 & 1 \\ x_1 & x_2 & x_3 \\ y_1 & y_2 & y_3 \end{vmatrix}}$$

OR

$$\boxed{\text{AREA } \triangle = \frac{1}{2}\left[(x_2 y_3 - x_3 y_2) - (x_1 y_3 - y_1 x_3) + (x_1 y_2 - y_1 x_2)\right]}$$

17. Division of a straight line in a Given Ratio $\lambda : \mu$.

$$\boxed{x = \frac{\lambda x_2 + x_1 \mu}{\lambda + \mu}} \quad \boxed{y = \frac{\lambda y_2 + y_1 \mu}{\lambda + \mu}}$$

WORKED EXAMPLE 12

A variable line passes through the point $(-2, 1)$ and meets the coordinate axes at A and B. Find the equation of the locus of the mid-point of AB.

Solution 12

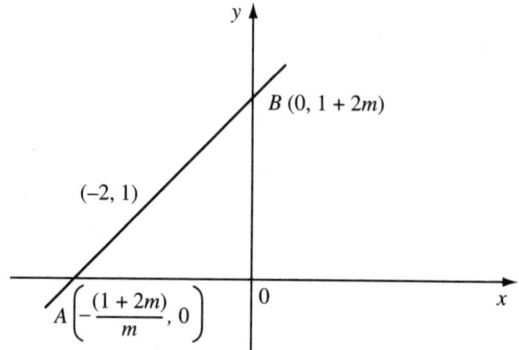

Fig. 6-I/24

Let $y = mx + c$ be the variable straight line, since it passes through the point $(-2, 1)$, $1 = -2m + c$, hence $c = 1 + 2m$.

The variable line is $y = mx + 1 + 2m$.

If $x = 0$, $y = 1 + 2m$; if $y = 0$, $x = -\dfrac{1 + 2m}{m}$.

If M is the mid-point, $M\left[-\dfrac{1}{2m}(1 + 2m), \dfrac{1}{2}(1 + 2m)\right]$.

Eliminating m from the equations

$$x = -\frac{1}{2m}(1 + 2m) = -\frac{1}{2m} - 1$$

$$y = \frac{1}{2}(1 + 2m)$$

$$2y = 1 + 2m \Rightarrow 2m = 2y - 1$$

$$x = -\frac{1}{2y - 1} - 1$$

$$x(2y - 1) + 2y - 1 + 1 = 0$$

$$\boxed{x(2y - 1) + 2y = 0}$$

Exercises 1

1. Determine the distance PQ if $P(3, 4)$, and $Q(-1, 2)$.

   Ans. 4.47 units.

2. Determine the distance AB if $A(-1, -2)$, and $B(-2, -5)$.

   Ans. 3.16 units.

3. Determine the distance RS if $R(2, 6)$, and $S(3, 5)$.

   Ans. 1.414 units.

4. Determine the perimeter of the triangle ABC, if $A(-3, -2)$, $B(1, 3)$ and $C(5, -1)$.

   Ans. 20.12 units.

5. Determine the area of the triangle of question 4.

   Ans. 18.0 square units.

6. Find the straight line that passes through the point $(2, 3)$ and has a gradient -1.

   Ans. $y = -x + 5$.

7. Find the straight line that passes through the point $(-1, -2)$ and has a gradient $\frac{2}{3}$.

   Ans. $3y - 2x + 4 = 0$.

8. Find the straight line that passes through the points $(1, 2)$ and $(0, 5)$.

   Ans. $y = -3x + 5$.

9. Find the straight line that passes through the sets of points $(-2, -3)$ and $(-4, -5)$.

   Ans. $y - x + 1 = 0$.

10. Determine the gradient of the line that passes through the points $(1, 3)$ and $(-2, 5)$.

    Ans. $-\frac{2}{3}$.

11. Determine the equation of the straight line that passes through the point $(-2, 4)$ and it is parallel to the line $3x + 4y - 1 = 0$.

    Ans. $4y + 3x - 10 = 0$.

12. Determine the equation of the straight line which is parallel to the line $-2x - y + 5 = 0$ and passes through the origin.

    Ans. $y = -2x$.

13. Write down the general forms of the lines:

    (i) $y = -3x - 5$

    (ii) $x + \frac{y}{2} = 1$.

    Ans. (i) $y + 3x + 5 = 0$ (ii) $2x + y - 2 = 0$.

14. Write down the gradient/intercept of the lines

    (i) $3x - 2y - 7 = 0$

    (ii) $\frac{x}{(-5)} + \frac{y}{2} = 1$.

    Ans. (i) $y = \frac{3}{2}x - \frac{7}{2}$, (ii) $y = \frac{2}{5}x + 2$.

15. Write down the intercept form of the lines:

    (i) $y = 3x - 1$

    (ii) $3x + 4y + 5 = 0$.

    Ans. (i) $\frac{x}{\frac{1}{3}} + \frac{y}{-1} = 1$ (ii) $\frac{x}{-\frac{5}{3}} + \frac{y}{-\frac{5}{4}} = 1$.

16. Determine the equation of the straight line which passes through the point $(0, -5)$ and it is perpendicular to $2x - y - 3 = 0$.

    Ans. $2y + x + 10 = 0$.

17. State the condition that two straight lines are parallel

    Ans. $m_1 = m_2$.

18. State the condition that two straight lines are perpendicular to each other.

    Ans. $m_1 m_2 = -1$.

19. Three straight lines of equations $3x + 4y - 1 = 0$, $2x - 3y + 5 = 0$ and $x - y + 6 = 0$ intersect each other. Find the perimeter of the triangle.

    Ans. 31.0 units.

20. Find the angle between the set of lines:

    (i) $3x + 4y - 7 = 0$
    $-x + 2y + 5 = 0$

    (ii) $-x + 3y - 2 = 0$
    $2x + y + 1 = 0$

    (iii) $4x - y - 1 = 0$
    $2x + 4y + 7 = 0$

    Ans. (i) $63.4°$ (ii) $81.9°$ (iii) $77.5°$)

21. Show that the triangle OAB is a right angled triangle, if $A(3, 0)$, $B(3, 4)$ and $O(0, 0)$.

22. Sketch the lines:
    (i) $y = 0$
    (ii) $x = 0$
    (iii) $x = -1$
    (iv) $x = 4$
    (v) $y = 3x$
    (vi) $y = 2x - 1$
    (vii) $y = x + 4$.

23. Find the perpendicular distance from the point $(2, 3)$ to the line $3x - 4y - 5 = 0$.
    Ans. $\dfrac{11}{5}$.

24. Find the perpendicular distance from the point $(-1, -4)$ to the line $x + y = 1$.
    Ans. $3\sqrt{2}$.

25. Determine the perpendicular distance from $(0, 0)$ to $x + y = 5$.
    Ans. $\dfrac{5\sqrt{2}}{2}$.

26. Sketch the straight lines:
    (i) $y = 2x - 3$
    (ii) $\dfrac{x}{1} + \dfrac{y}{-2} = 1$
    (iii) $2x + 3y + 4 = 0$.

27. Sketch the graphs:
    (i) $y = -2x$ (ii) $y = 3x$ (iii) $y = x$.

28. Sketch the following straight lines:
    (i) $3y = 2x$
    (ii) $y = -x + 5$
    (iii) $y = 2x + 1$
    (iv) $y = 0$
    (v) $x = 0$
    (vi) $y = 3$
    (vii) $y = -2$
    (viii) $y = -x$
    (ix) $y = 5x$
    (x) $y = 3x$.

29. Show that the lines $y = x$ and $y + x = 0$ are perpendicular.

30. Find the mid-points, M and N, between the pair of points $A(-5, 5)$, $B(3, 7)$; and $D(-1, 5)$, $C(7, 7)$, hence find the mid-point of MN.
    Ans. $(1,6)$.

31. Show that the points $A(3, 1)$, $B(3, 7)$, $C(20, 7)$ and $D(20, 1)$ form the vertices of a rectangle.

32. Show that the points $A(-5, 5)$, $B(3, 7)$, $C(7, 7)$ and $D(-1, 5)$ form the vertices of a parallelogram.

33. Determine the coordinates of the mid-points between the pairs of points of the following:
    (i) $(-3, -4)$, $(3, 4)$
    (ii) $(-1, -3)$, $(1, 3)$
    (iii) $(-2, 5)$, $(3, -4)$
    (iv) $(0, 5)$, $(4, 0)$
    (v) $(3, 4)$, $(7, 9)$

    Ans. (i) $(0, 0)$ (ii) $(0, 0)$ (iii) $\left(\dfrac{1}{2}, \dfrac{1}{2}\right)$
    (iv) $(2, 2.5)$ (v) $(5, 6.5)$.

34. Determine the distances between the pairs of points of the question 33.
    Ans. (i) 10 (ii) $\sqrt{40}$ (iii) $\sqrt{106}$
    (iv) $\sqrt{41}$ (v) $\sqrt{41}$.

35. Find the equation of the line passing through the point $(1, 2)$ with gradient equal to -1.
    Ans. $y = -x + 3$.

36. Find the equation of the line passing through $P_1(-2, 3)$ and $P_2(3, 5)$.
    Ans. $5y - 2x - 19 = 0$.

37. Find the equation of the line passing through the point $(-2, -3)$ with gradient equal to $\frac{1}{2}$.
    Ans. $2y - x + 4 = 0$.

38. Find the equation of the line passing through the points $P_1(1, -2)$ and $P_2(-3, -4)$.
    Ans. $2y - x + 5 = 0$.

39. Find the perpendicular distance from the point $(3, 4)$ to the line $y = -x$
    Ans. $\dfrac{7}{\sqrt{2}}$.

40. Find the perpendicular distances from the point $(0, -2)$ to the lines $x + y = 3$ and $x + y + 5 = 0$.

    Ans. $\dfrac{5\sqrt{2}}{2}, \dfrac{3\sqrt{2}}{2}$.

41. Show that the acute angle between the lines $3x + y - 7 = 0$ and $2x - y + 5 = 0$ is $45°$.

42. Show that the obtuse angle between the lines $y = 7 - 3x$ and $y = 2x + 5$ is $135°$.

43. The straight lines $ax - y + 5 = 0$ and $2ax + y + 8 = 0$ intersect at an acute angle whose tangent is $\frac{1}{2}$. Find the values of a.

    Ans. $a = 0.158, a = -3.158$.

44. The straight lines $2x + 4y = 3$, $ax - y - 5 = 0$ are inclined to each other at an angle of $75°$. Find the value of a.

    Ans. $a = 1.128$.

45. ABC is a triangle in which the vertices are $A(2, -3)$, $B(-2, 5)$, $C(2, 6)$. Determine the gradients of the sides AB, BC, and AC. Hence calculate the angles of the triangle, M is the mid-point of AB, determine the gradient of CM and its angle with AB.

    Ans. $-2, \dfrac{1}{4}, \infty, 26.6°, 75.9°, 77.5°, M(0, 1), 48.4°$.

46. Calculate the acute angle (in degrees and minutes) between the lines.

    (a) $2x + y = 1, 3x + 4y = 5$
    (b) $x + 2y = 3, 3x + 2y = 1$
    (c) $-x + 5y = 7, 2x + 3y = 4$
    (d) $5x + 4y = 3, -2x + 3y = -4$.

    Ans. (a) $26°34'$ (b) $\theta = 29°45'$ (c) $45°$ (d) $85°2'$.

47. Prove that the product of the gradients of two perpendicular lines is equal to -1.

48. Show that the angle Ψ between the lines $y = m_1 x + c_1$ and $y = m_2 x + c_2$ is given by $\tan \Psi = \dfrac{m_1 - m_2}{1 + m_1 m_2}$.

49. Show that the tangent of the acute angle between the lines $y = -\frac{1}{2}x + \frac{5}{2}$ and $y = \frac{2}{3}x + \frac{7}{3}$ is equal to $\frac{7}{4}$.

50. Find the equations of the straight lines through the point $(-2, 1)$ parallel and perpendicular to $\frac{x}{1} + \frac{y}{-2} = 1$.

    Ans. $y = 2x + 5, y = -\dfrac{1}{2}x$.

51. Find the equations of the straight lines through the point $(3, 2)$ parallel and perpendicular to $\frac{x}{-1} + \frac{y}{3} = 1$.

    Ans. $y = 3x - 7, 3y + x = 9$.

52. Find the perpendicular length from the point $(1, -5)$ to the line $\frac{x}{3} + \frac{y}{2} = 1$.

    Ans. 5.27 units.

53. Find the perpendicular length from the point $(-3, 2)$ to the line $3x - y - 3 = 0$.

    Ans. 4.43 units.

54. Find the acute angle between the lines $2x + 5y = 1$ and $x - 3y - 4 = 0$.

    Ans. $40.2°$.

55. Find the acute angle between the lines $x + y - 5 = 0$ and $3x - 5y + 2 = 0$.

    Ans. $76°$.

56. Find the equation of the straight line joining the origin to the point of intersection of $2x + 5y = 1$ and $3x - 5y + 2 = 0$.

    Ans. $y = -\dfrac{7}{5}x$.

57. Find the equation of the straight line joining the point $(-1, 2)$ to the point of intersection of $x - 3y - 4 = 0$ and $3x - 5y + 2 = 0$.

    Ans. $y - x - 3 = 0$.

58. Find the equation of the straight line through the point of intersection $x - 3y - 4 = 0$ and $3x - y - 3 = 0$ and parallel to $y + 3x + 7 = 0$.

    Ans. $4y + 12x = 3$.

59. Find the equation of the straight line through the point of intersection of $3x - y - 3 = 0$ and $3x - 5y + 2 = 0$ and perpendicular to $-2x + y + 2 = 0$.

    Ans. $24y + 12x = 47$.

60. The vertex A of the triangle ABC is the point $(1, -3)$ and the equation of the side BC is $x - 3y + 11 = 0$. The mid-point of BC has x-coordinate 1 and the area of the triangle ABC is 21 square units. Find the coordinates of B and C.

    Ans. $B(-2, 3), C(4, 5)$.

61. The sides of a triangle ABC are $\sqrt{45}, \sqrt{73}, \sqrt{40}$ respectively. Determine the angles of the triangle and the area of ABC.

    Ans. $51°, 81.9°, 47.1°$; 21 square units.

62. The vertices of the triangle ABC are given $A(1, -3)$, $B(-2, 3)$, $C(4, 5)$. Calculate the gradient of the lines AB, BC and AC and hence determine the angles of the triangle.

   Ans. $-2, \dfrac{1}{3}, \dfrac{8}{3}$; $A = 51°, B = 81.9°, C = 47.1°$.

63. Determine the heights of the triangle with coordinates $A(1, -3)$, $B(-2, 3)$ and $C(4, 5)$. Hence find the area of the triangle.

   Ans. $p = \dfrac{21}{\sqrt{10}}, q = \dfrac{42}{\sqrt{73}}, r = \dfrac{14}{\sqrt{5}}$; 21 square units.

64. The coordinates of a triangle PQR are $(1, -3)$, $(-2, 3)$, and $C(2, 5)$. Find the mid-points of the sides and hence the coordinates of the meet of the medians. Find also the coordinates of the meet of the heights.

   Ans. $\left(-\dfrac{1}{2}, 0\right), (0, 4), \left(\dfrac{3}{2}, 1\right)$; $\left(\dfrac{1}{3}, \dfrac{5}{3}\right)$; $(-2, 3)$.

65. Find the perpendicular distance from the known point (x_1, y_1) from the known line $ax + by + c = 0$.

66. Find the equations of the bisectors of the angles between the lines $y = 0$ and $x = 0$.

67. Find the equations of the bisectors of the angles between the lines $y = x$ and $y = -x$.

68. Find the equations of the bisectors of the angles between the lines $3x - 4y - 2 = 0$ and $5x + 12y - 1 = 0$.

69. Determine the equations of the bisectors of the angles between the lines $l_1 : a_1x + b_1y + c_1 = 0$ and $l_2 : a_2x + b_2y + c_2 = 0$.

70. Find the equation of the straight line passing through the point of intersection of two given straight lines $a_1x + b_1y + c_1 = 0$ and $a_2x + b_2y + c_2 = 0$.

71. Find the equation of the straight line passing through the point of intersection of the lines $x + y = -3$, $2x - y = 6$.

CONIC SECTIONS

2

The Circle

A horizontal plane intersects a right circular cone into a circle.

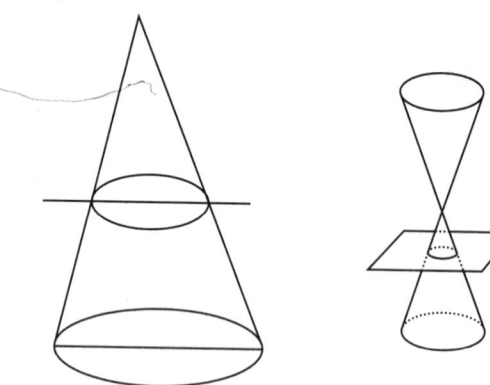

Fig. 6-1/25

A circle with its centre at the origin has an

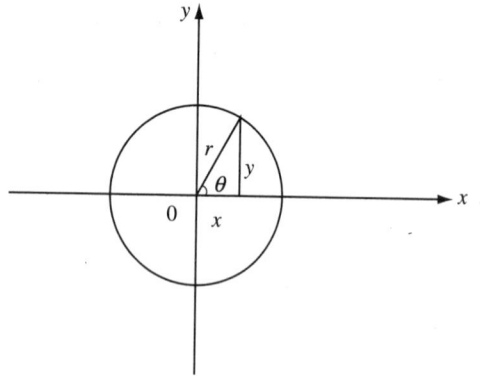

Fig. 6-1/26

equation $\boxed{x^2 + y^2 = r^2}$ where (x, y) is any point on the circumference.

$$\sin \theta = \frac{y}{r} \qquad \cos \theta = \frac{x}{r}$$

$$\sin^2 \theta + \cos^2 \theta = 1 = \frac{y^2}{r^2} + \frac{x^2}{r^2}$$

$$x^2 + y^2 = r^2.$$

The parametric equations are $x = r \cos \theta$, and $y = r \sin \theta$.

If the centre of the circle has coordinates $(-g, -f)$, the equation of the circle is

$$(x + g)^2 + (y + f)^2 = r^2$$

$$x^2 + 2gx + g^2 + y^2 + 2fy + f^2 - r^2 = 0$$

$$x^2 + 2gx + y^2 + 2fy + g^2 + f^2 - r^2 = 0$$

where $c = g^2 + f^2 - r^2$ is a constant

$$\boxed{x^2 + 2gx + y^2 + 2fy + c = 0} \qquad \ldots (1)$$

This is the general equation of the circle with centre $c(-g, -f)$ and radius $r = \sqrt{g^2 + f^2 - c}$.

WORKED EXAMPLE 13

Determine the equation of the circle which passes through three sets of points $(-2, -3)$, $(1, 3)$, $(2, 1)$ hence find the coordinates of the centre and the radius.

Solution 13

Applying the general equation of the circle (1), we have

For $x = -2, y = -3$

$$(-2)^2 + 2g(-2) + (-3)^2 + 2f(-3) + c = 0$$

$$4 - 4g + 9 - 6f + c = 0$$

$$\boxed{4g + 6f - c = 13} \quad \ldots (2)$$

For $x = 1, y = 3$

$$1^2 + 2g(1) + 3^2 + 2f(3) + c = 0$$

$$\boxed{2g + 6f + c = -10} \quad \ldots (3)$$

For $x = 2, y = 1$

$$2^2 + 2g(2) + 1^2 + 2f(1) + c = 0$$

$$\boxed{4g + 2f + c = -5} \quad \ldots (4)$$

(4) − (2)

$$\boxed{-4f + 2c = -18} \quad \ldots (5)$$

(4) − 2(3)

$$\boxed{-10f - c = 15} \quad \ldots (6)$$

(5) + 2(6)

$$-24f = 12 \quad \boxed{f = -\frac{1}{2}}$$

$$2c = -18 + 4\left(-\frac{1}{2}\right) = -18 - 2 \quad \boxed{c = -10}$$

$$4g + 6f - c = 13$$

$$4g + 6\left(-\frac{1}{2}\right) + 10 = 13$$

$$4g = 6 \quad \boxed{g = \frac{3}{2}}$$

$$x^2 + 2\left(\frac{3}{2}\right)x + y^2 + 2\left(-\frac{1}{2}\right)y - 10 = 0$$

$$\boxed{x^2 + 3x + y^2 - y - 10 = 0}$$

$$\left(x + \frac{3}{2}\right)^2 - \frac{9}{4} + \left(y - \frac{1}{2}\right)^2 - \frac{1}{4} - 10 = 0$$

$$\left(x + \frac{3}{2}\right)^2 + \left(y - \frac{1}{2}\right)^2 = 10 + \frac{1}{4} + \frac{9}{4} = \frac{50}{4} = \frac{25}{2}$$

$$\boxed{C\left(-\frac{3}{2}, \frac{1}{2}\right)} \quad \boxed{r = \frac{5}{\sqrt{2}}}.$$

WORKED EXAMPLE 14

Find the equation of a circle whose centre is $c(-1, -2)$ and the radius $r = 2$.

Solution 14

$$(x + 1)^2 + (y + 2)^2 = 2^2$$

$$x^2 + 2x + 1 + y^2 + 4y + 4 - 4 = 0$$

$$\boxed{x^2 + 2x + y^2 + 4y + 1 = 0}$$

WORKED EXAMPLE 15

Determine the centre and the radius of the circle $x^2 - 18x + y^2 - 6y + 65 = 0$.

Solution 15

$$[x^2 - 18x] + [y^2 - 6y] + 65 = 0$$

$$[(x - 9)^2 - 81] + [(y - 3)^2 - 9] + 65 = 0$$

by completing the squares in the square brackets.

$$(x - 9)^2 + (y - 3)^2 = 5^2.$$

The centre and radius are given respectively $C(9, 3)$ and $r = 5$.

WORKED EXAMPLE 16

Find the condition that the straight line $y = mx + 2$ is tangent to the circle $x^2 + x + y^2 + 2y + 1 = 0$. Determine the coordinates of the point of contact.

Solution 16

$$x^2 + x + (mx + 2)^2 + 2(mx + 2) + 1 = 0$$

$$x^2 + x + m^2x^2 + 4mx + 4 + 2mx + 4 + 1 = 0$$

$$(m^2 + 1)x^2 + x(1 + 6m) + 9 = 0$$

where $D = b^2 - 4ac = 0$

$$(1 + 6m)^2 - 36(m^2 + 1) = 0$$

$$1 + 12m + 36m^2 - 36m^2 - 36 = 0$$

$$12m = 35$$

$$m = \frac{35}{12}.$$

The straight line is $y = \frac{35}{12}x + 2$

or $12y - 35x - 24 = 0$

$$x = \frac{-(1+6m) \pm \sqrt{(1+6m)^2 - 36(m^2+1)}}{2(m^2+1)}$$

$$x = \frac{-(1+6m)}{2(1+m^2)}$$

$$= -\frac{1 + 6 \times \frac{35}{12}}{2\left(1 + \frac{35^2}{12^2}\right)}$$

$$= -\frac{18.5}{19.014} = -0.973$$

$$y = \frac{35}{12}x + 2$$

$$= \frac{35}{12}(-0.973) + 2$$

$y = -0.84$.

The point of contact $(-0.973, -0.84)$.

Orthogonal Circles

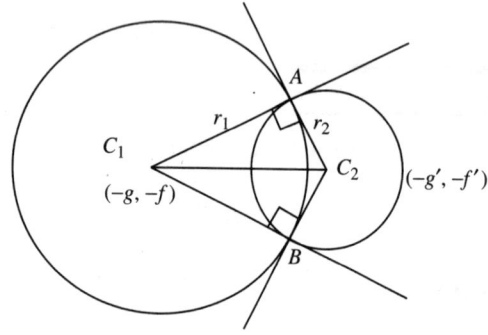

Fig. 6-1/27

Two circles intersect orthogonally, that is, they intersect at right angles at A and at B.

$(C_1C_2)^2 = (AC_1)^2 + (AC_2)^2$

$(C_1C_2)^2 = (BC_1)^2 + (BC_2)^2$

C_1A is the tangent at A of the circle with radius C_2A and C_2B is the tangent at B of the circle with radius C_1B.

The condition that $x^2 + y^2 + 2gx + 2fy + c = 0$ and $x^2 + y^2 + 2g'x + 2f'y + c = 0$ should cut orthogonally.

The centres of the circle are $C_1(-g, -f)$, $C_2(-g', -f')$ respectively ant their radii are $\sqrt{g^2 + f^2 - c}$, $\sqrt{g'^2 + f'^2 - c'}$ respectively $\boxed{r_1^2 + r_2^2 = C_1C_2^2}$

$$[-g - (-g')]^2 + [-f - (-f')]^2$$
$$= g^2 + f^2 - c + g'^2 + f'^2 - c'$$

$$g^2 + g'^2 - 2gg' + f^2 + f'^2 - 2ff'$$
$$= g^2 + f^2 - c + g'^2 + f'^2 - c'$$

$$\boxed{2gg' + 2ff' = c + c'}$$

The condition that the circles are orthogonal.

WORKED EXAMPLE 17

Determine whether the two circles are orthogonal.

(i) $(x-2)^2 + (y+4)^2 = 9$

(ii) $(x+3)^2 + (y-2)^2 = (3\sqrt{5})^2$.

Solution 17

(i) $g = 2, f = -4$

(ii) $g' = -3, f' = 2$

$$2gg' + 2ff' = 2 \times 2(-3) + 2 \times (-4)2$$
$$= -28$$

$$r_1 = \sqrt{g^2 + f^2 - c} = 3,$$

$$(-2)^2 + (-4)^2 - c = 3^2$$

$$c = 4 + 16 - 9 = 11$$

$$r_2 = \sqrt{g'^2 + f'^2 - c'}$$

$$= \sqrt{(-3^2) + (2)^2 - c'}$$

$$= 3\sqrt{5}$$

$$9 + 4 - 45 = c' = -32$$

$$c + c' = 11 - 32$$

$$= -21.$$

The circles are not orthogonal.

Worked Example 18

Determine which of the pairs of circles are orthogonal.

(a) (i) $(x+3)^2 + (y-4)^2 = 20$,

(ii) $(x+4)^2 + (y+3)^2 = 20$

(b) (i) $x^2 + y^2 = 1$,

(ii) $(x-3)^2 + (y+4)^2 = 24$.

Solution 18

(a) (i) $x^2 + 6x + 9 + y^2 - 8y + 16 - 20 = 0$

or $x^2 + 6x + y^2 - 8y + 5 = 0$

$g = -3, f = 4, c = 5$

(ii) $x^2 + 8x + 16 + y^2 + 6y + 9 - 20 = 0$

or $x^2 + 8x + y^2 + 6y + 5 = 0$

$g' = -4, f' = -3, c' = 5$

$2gg' + 2ff' = 2(-3)(-4) + 2(4)(-3) = 0$

$c + c' = 10$

(i) and (ii) are not orthogonal.

(b) (i) $x^2 + y^2 - 1 = 0$

(ii) $x^2 - 6x + 9 + y^2 + 8y + 16 - 24 = 0$

From

(i) $g = 0, f = 0, c = -1$.

From

(ii) $g' = 3, f' = -4, c' = 1$

$2gg' + 2ff' = 0$

$c + c' = 0$.

The circles are orthogonal, since

$2gg' + 2ff' = c + c'$.

Worked Example 19

Show that the circle $x^2 + y^2 + 6x - 10y - 2 = 0$ is orthogonal to both circles

$x^2 + y^2 + 10x + 3y + 17 = 0$

$x^2 + y^2 - 7x - 5y + 6 = 0$.

Solution 19

$x^2 + y^2 + 6x - 10y - 2 = 0$

completing the squares

$(x+3)^2 - 9 + (y-5)^2 - 25 - 2 = 0$

$(x+3)^2 + (y-5)^2 = 36$

$g = -3, f = 5, c = -2$

$2gg' + 2ff' = c + c'$ condition for two circles to be orthogonal

$x^2 + y^2 + 10x + 3y + 17 = 0$

$(x+5)^2 - 25 + \left(y + \frac{3}{2}\right)^2 - \frac{9}{4} + 17 = 0$

$g' = -5, f' = -\frac{3}{2}, c' = 17$

$2gg' + 2ff' = 2(-3)(-5) + 2(5)\left(-\frac{3}{2}\right) = 15$

$c + c' = -2 + 17 = 15$

$x^2 + y^2 - 7x - 5y + 6 = 0$

$\left(x - \frac{7}{2}\right)^2 - \frac{49}{4} + \left(y - \frac{5}{2}\right)^2 - \frac{25}{4} + 6 = 0$

$g' = \frac{7}{2}, f' = \frac{5}{2}, c' = 6$

$2gg' + 2ff' = 2(-3)\left(\frac{7}{2}\right) + 2(5)\left(\frac{5}{2}\right)$

$= -21 + 25 = 4$

$= c + c' = -2 + 6 = 4$.

Parametric Equation of a Circle

The coordinates of a curve are given $x = r\cos\theta \ldots (1)$ and $y = r\sin\theta \ldots (2)$. Determine the locus of this curve with these parameters.

From (1) $\cos\theta = \frac{x}{r}$ and from (2) $\sin\theta = \frac{y}{r}$. Squaring both sides of these equations we have $\cos^2\theta = \frac{x^2}{y^2}$ and $\sin^2\theta = \frac{y^2}{r^2}$ and adding we have

$\cos^2\theta + \sin^2\theta = \frac{x^2}{r^2} + \frac{y^2}{r^2} = 1$.

Therefore $\boxed{x^2 + y^2 = r^2}$ is the cartesian equation of the circle with $C(0, 0)$ and radius r. The parametric equations of this circle are $x = r\cos\theta$ and $y = r\sin\theta$. If this circle has centre (a, b) and radius r, the corresponding equations are $x = a + r\cos\theta$ and $y = b + r\sin\theta$. The locus of the curve is found by eliminating the parameter θ, $\cos\theta = \frac{x-a}{r}$ and $\sin\theta = \frac{y-b}{r}$.

$$\cos^2\theta + \sin^2\theta = 1 = \left(\frac{x-a}{r}\right)^2 + \left(\frac{y-b}{r}\right)^2$$

$$\boxed{(x-a)^2 + (y-b)^2 = r^2}$$

If the general case of the circle $C(-g, -f)$ and the equation is

$$(x+g)^2 + (y+f)^2 = r^2$$

$$x^2 + 2xg + g^2 = y^2 + 2fy + f^2 = r^2$$

$$x^2 + y^2 + 2xg + 2fy + g^2 + f^2 - r^2 = 0$$

if $c = g^2 + f^2 - r^2$ = constant

$$\boxed{x^2 + y^2 + 2xg + 2fy + c = 0}$$

the general equation of the circle.

WORKED EXAMPLE 20

Determine the Cartesian equations of the following curves:

(i) $x = 3\cos\theta$, $y = 3\sin\theta$

(ii) $x = \cos\theta$, $y = \sin\theta$

(iii) $x = 1 + 2\cos\theta$, $y = 3 + 2\sin\theta$

(iv) $x = -2 + 5\cos\theta$, $y = 5 + 5\sin\theta$.

Solution 20

(i) $x = 3\cos\theta$, $\cos\theta = \frac{x}{3}$; $y = 3\sin\theta$, $\sin\theta = \frac{y}{3}$,

To eliminate the parameter θ we use $\cos^2\theta + \sin^2\theta = 1$

$$\cos^2\theta + \sin^2\theta = 1 = \left(\frac{x}{3}\right)^2 + \left(\frac{y}{3}\right)^2, \text{ therefore}$$

$$\boxed{x^2 + y^2 = 3^2}$$

(ii) $x = \cos\theta$, $\cos\theta = x$; $y = \sin\theta$, $\sin\theta = y$.

$$\cos^2\theta + \sin^2\theta = 1 = x^2 + y^2$$

$$\boxed{x^2 + y^2 = 1}$$

(iii) $x = 1 + 2\cos\theta$, $\cos\theta = \frac{x-1}{2}$;

$y = 3 + 2\sin\theta$, $\sin\theta = \frac{y-3}{2}$.

$$\cos^2\theta + \sin^2\theta = \left(\frac{x-1}{2}\right)^2 + \left(\frac{y-3}{2}\right)^2 = 1,$$

$$\boxed{(x-1)^2 + (y-3)^2 = 2^2}$$

(iv) $x = 2 + 5\cos\theta$, $\cos\theta = \frac{x+2}{5}$;

$y = 5 + 5\sin\theta$, $\sin\theta = \frac{y-5}{5}$.

$$\cos^2\theta + \sin^2\theta = 1 = \left(\frac{x+2}{5}\right)^2 + \left(\frac{y-5}{5}\right)^2$$

$$\boxed{(x+2)^2 + (y-5)^2 = 5^2}$$

WORKED EXAMPLE 21

Find the parametric equation of the following curves

(i) $x^2 + y^2 = 2$

(ii) $x^2 + y^2 - 4x = 6y - 12 = 0$

(iii) $x^2 + y^2 + 2gx + 2fy + c = 0$.

Solution 21

(i) $x^2 + y^2 = (\sqrt{2})^2$

$x = \sqrt{2}\cos\theta$ and $y = \sqrt{2}\sin\theta$ are the parametric equations.

(ii) $x^2 + y^2 - 4x - 6y - 12 = 0$, completing the squares

$(x-2)^2 - 4 + (y-3)^2 - 9 - 12 = 0$,

$(x-2)^2 + (y-3)^2 = 5^2$

$x = 2 + 5\cos\theta$, $y = 3 + 5\sin\theta$ are the parametric equations.

(iii) $x^2 + y^2 + 2gx + 2fy + c = 0$, completing the squares

$(x+g)^2 - g^2 + (y+f)2 - f^2 + c = 0$

$(x+g)^2 + (y+f^2) = \left(\sqrt{g^2 + f^2 - c^2}\right)$

$x = -g + \sqrt{g^2 + f^2 - c}\cos\theta$,

$y = -f + \sqrt{g^2 + f^2 - c}\sin\theta$ are the parametric equations.

Exercises 2

1. The parametric equations of a circle are given by $x = 2\cos\theta$ and $y = 2\sin\theta$ determine the equation of the circle.

2. The parametric equations of a circle are given by $x = 5\cos\theta$ and $y = 5\sin\theta$ determine the equation of the circle.

3. The centre of a circle is $(0, 0)$ and its radius is 7, find the equation of the circle.

4. The equation of a circle is $x^2 + y^2 = 2^2$, determine the centre and its radius.

5. If the centre of a circle is $C(-g, -f)$ and the radius $\sqrt{g^2 + f^2 - c}$ find the equation of the circle.

6. A circle passes through the three points $(0, 0)$, $(-1, 5)$ and $(3, 4)$ determine the equation of the circle.

7. A circle passes through the points $(0, 5)$, $(7, 0)$ and $(-4, -5)$. Determine the equation of the circle.

8. A circle has a centre $(0, 6)$ and radius 4, find the equation of the circle.

9. Determine the equation of the circle which has a centre $(3, 2)$ and passes through the point $(0, 5)$.

10. Find the equations of the circles with
    (i) $C(0, 3), r = 3$
    (ii) $C(-1, 2), r = 2$
    (iii) $C(2, -3), r = 1$

11. Determine the centre and the radius of the circle $x^2 + x + y^2 + 2y + 1 = 0$.

12. Determine the centre and the radius of the circle $4x^2 + 5x + 4y^2 + 3y - \dfrac{111}{8} = 0$.

13. Find the points of intersection of the two circles: $x^2 - 4x + y^2 - 6y + 12 = 0$ and $x^2 - 2x + y^2 - 2y - 2 = 0$.

14. Determine the coordinates of the centre and the radius of a circle which passes through the points $A(1, 5)$, $B(3, 1)$, $C(7, 3)$.

15. Show that the point $(3\cos\theta, 3\sin\theta)$ lies on the circle $x^2 + y^2 = 9$ for all values of θ

16. Show that the point $(2 + 5\cos\theta, -3 + 5\sin\theta)$ lies on the circle
$$x^2 + y^2 + 6y - 4x - 12 = 0.$$

17. Show that the point $(-5 + \cos\theta, -1 + \sin\theta)$ lies on the circle
$$x^2 + 10x + y^2 + 2y + 25 = 0.$$

18. Find the equation of the circle of centre $(-3, -4)$ which passes through the point $(2, -3)$.

19. Find the equations of the circles with the following centres and radii:
    (a) $(-2, 3); 5$
    (b) $(-2, -2); 2$
    (c) $\left(\sqrt{5}, \sqrt{3}\right); \sqrt{2}$

20. Determine the equation of the circle whose centre is $(-g, -f)$ and its radius $r = \sqrt{g^2 + f^2 - c}$.

21. Determine the condition that the straight line $y = mx + k$ is tangent to the circle
$$x^2 + y^2 + 2xg + 2yf + c = 0.$$

22. Find the equation of the circle passing through the points $(1, 2), (-2, 3)$, and $(1, -5)$.

23. Find the equation of the circle passing through the points $(3, 2), (-3, 2)$, and $(0, 0)$.

24. Write down the equation of the circle whose centre is the origin and the radius is 5.

25. Write down the equation of the circle whose centre is $(-3, 4)$ and the radius is 3.

26. If the centre of a circle is $(-g, -f)$ and its radius is $\sqrt{g^2 + f^2 - c}$, write down the general equation of the circle.

27. Repeat question 26, if $C(2, 3)$ and the radius is 1.

28. Determine the equation of the tangent at (x_1, y_1) to the circle
$$x^2 + y^2 + 2gx + 2fy + c = 0.$$

29. Show that the equations of the tangents of gradients m to $x^2 + y^2 = r^2$ are $y = mx \pm r\sqrt{1 + m^2}$

30. Show that the square of the length of the tangent from $A(x_1, y_1)$ to the circle

$$x^2 + y^2 + 2gx + 2fy + c = 0 \text{ is}$$

$$x_1^2 + y_1^2 + 2gx_1 + 2fy_1 + c$$

31. Write down the equations of the tangents at the given points to the following circles:

    (i) $x^2 + y^2 - 3x + 4y + 6 = 0$ $(2, -2)$

    (ii) $x^2 + y^2 + x + y = 0$, $(-1, -1)$.

32. Find the equation of the circle which passes through the origin and the points $(2, 8)$ and $(4, -2)$.

33. Find the equation of the circle which has the line segment joining the points $(-3, 4)$ and $(2, -3)$ as diameter.

34. Three points A, B and C have coordinates $(2, 4)$, $(8, -2)$ and $(6, 2)$ respectively. Determine the equation of the circle which passes through these points, hence determine the coordinates of the centre and the radius.

35.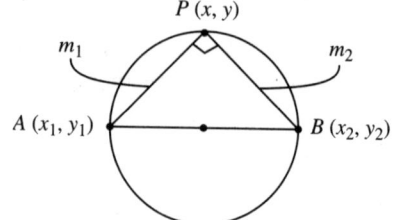

Fig. 6-I/28

Determine the equation of the circle knowing the coordinates of the ends of a diameter.

36. Determine the coordinates of the centre C and the radius of the circle with equation $x^2 + y^2 - 2y - 4x - 4 = 0$. The circle cuts the y-axis at the points A and B. Calculate

    (i) the exact area of the $\triangle ABC$.

    (ii) the area of the segment enclosed by the chord AB and the minor length of arc AB. The circle intersects the x-axis at the points D and E. Calculate the area of the quadrilateral $ADBE$.

3

Parabolas

Equations of the Parabola

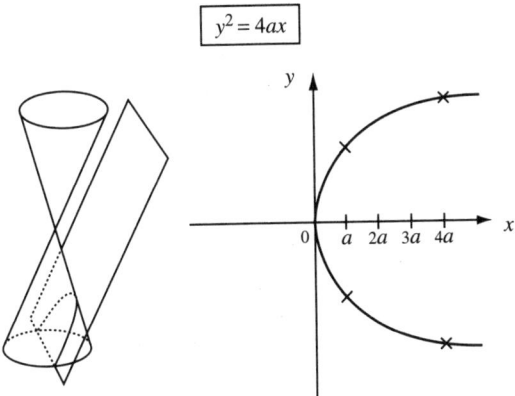

Fig. 6-I/29

If $x = 0, y = 0$; if $x = a, y = \pm 2a$; if $x = 4a$, $y^2 = 16a^2, y = \pm 4a$.

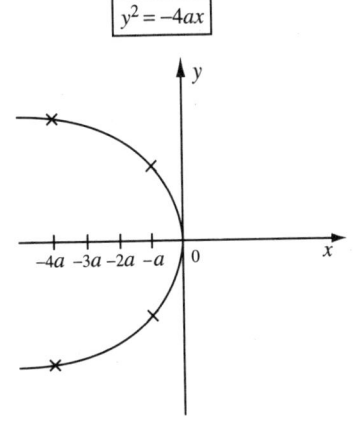

Fig. 6-I/30

If $x = 0, y = 0$; $x = -a, y = \pm 2a$; if $x = -4a$, $y^2 = 16a^2, y = \pm 4a$.

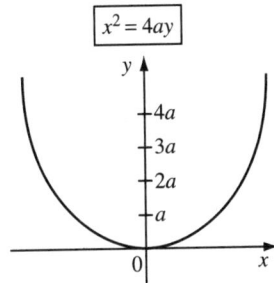

Fig. 6-I/31

If $y = 0, x = 0$; if $y = a, x = \pm 2a$; if $y = 4a$, $x = \pm 4a$.

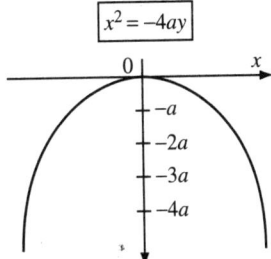

Fig. 6-I/32

If $y = 0, x = 0$; if $y = -a, x = \pm 2a$; if $y = -4a$, $x = \pm 4a$.

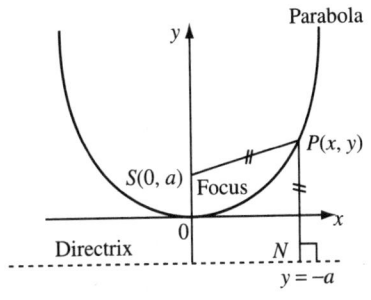

Fig. 6-I/33 Definition of parabola

25

The locus of a point $P(x, y)$ which moves so that it is equidistant from the fixed focus point $S(0, a)$ and the fixed line $y = -a$, is a parabola with equation $x^2 = 4ay$.

$$PS = PN$$

$$PN^2 = (y + a)^2,$$

$$PS^2 = (y - a)^2 + x^2$$

$$(y - a)^2 + x^2 = (y + a)^2$$

$$y^2 - 2ay + a^2 + x^2 = y^2 + 2ay + a^2$$

$$\boxed{x^2 = 4ay}$$

Parameters

It is possible to find the x and y coordinates of a point in terms of a variable (say t). The use of a third variable, called a parameter, is often made to simplify a proof. Let $x = 2at$ and $y = at^2$, eliminate t from these two equations

$$t = \frac{x}{2a}, \; y = at^2 = a\left(\frac{x}{2a}\right)^2 = \frac{ax^2}{4a^2}$$

$$\boxed{x^2 = 4ay}$$

the locus of all the points x and y is a parabola given by $x^2 = 4ay$ and it is independent of t.

Consider two points $P(x_1, y_1)$ and $Q(x_2, y_2)$, with parametric coordinates $P(2ap, ap^2)$ and $Q(2aq, aq^2)$ respectively as shown in Fig. 6-I/34.

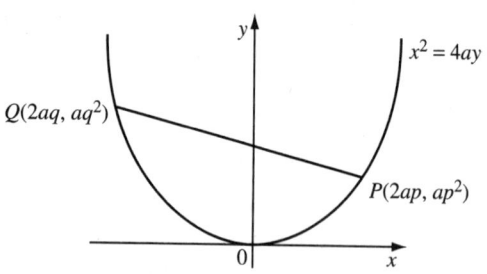

Fig. 6-I/34 Chord PQ

Gradient of Chord PQ

$$m = \frac{y_2 - y_1}{x_2 - x_1} = \frac{aq^2 - ap^2}{2aq - 2ap}$$

$$= \frac{a(q - p)(q + p)}{2a(q - p)} = \frac{q + p}{2}.$$

Equation of Chord

$$\frac{y - y_1}{x - x_1} = \frac{p + q}{2}$$

$$y - ap^2 = \frac{p + q}{2}(x - 2ap)$$

$$2y - 2ap^2 = (p + q)x - 2ap(p + q)$$

$$2y - 2ap^2 = px + qx - 2ap^2 - 2apq$$

$$2y - (p + q)x + 2apq = 0.$$

Equation of Tangent to $x^2 = 4ay$

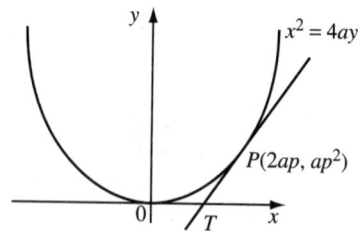

Fig. 6-I/35 Tangent at P.

The gradient of tangent at $P(2ap, ap^2)$.
$x^2 = 4ay$, differentiating with respect to y

$$2x \frac{dx}{dy} = 4a,$$

$$\frac{dx}{dy} = \frac{4a}{2x} = \frac{2a}{x}, \text{ or }$$

$$\boxed{\frac{dy}{dx} = \frac{x}{2a}}$$

At $P(2ap, ap^2)$, $\frac{dy}{dx} = \frac{2ap}{2a} = p$, the gradient of the tangent at P.

The equation of tangent at P is $y = mx + c$ or $y = px + c$, this passes through the point P, $ap^2 = p(2ap) + c$, $c = ap^2 - 2ap^2 = -ap^2$ therefore

$$\boxed{y = px - ap^2}$$

This equation can be obtained from the equation of the chord PQ by letting q to approach p.

$y - \frac{1}{2}(p+q)x + apq = 0$ the equation of the chord or

$$y = \frac{1}{2}(p+q)x - apq$$

as $q \to p$ $\quad y = \frac{1}{2}(p+p)x - app$

$$\boxed{y = px - ap^2}$$

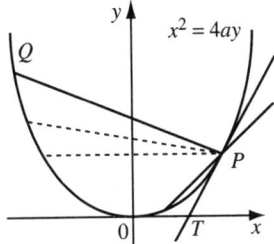

Fig. 6-I/36 Tangent at P

Equation of Tangent at $P(x_1, y_1)$ on $x^2 = 4ay$

$x^2 = 4ay$, $\frac{dy}{dx} = \frac{x}{2a}$, at $P(x_1, y_1)$, $\frac{dy}{dx} = \frac{x_1}{2a}$,

the gradient of the tangent at P.

$y = mx + c$, $y = \frac{x_1}{2a}x + c$, this passes through the point

P, $y_1 = \frac{x_1^2}{2a} + c$,

$c = y_1 - \frac{x_1^2}{2a}$ and therefore $y = \frac{x_1}{2a}x + y_1 - \frac{x_1^2}{2a}$,

$\boxed{xx_1 = 2ay + 2ay_1}$ since $x_1^2 = 4ay_1$

Equation of Normal at $P(2ap, ap^2)$

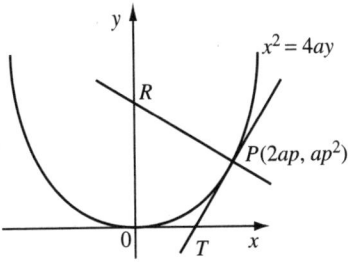

Fig. 6-I/37 Normal at $P(2ap, ap^2)$

$x^2 = 4ay$, $\frac{dy}{dx} = p$ at $P(2ap, ap^2)$, the gradient of the normal is $-\frac{1}{p}$ since $m_1 m_2 = -1$, $m_1 = p$ = gradient of tangent, m_2 = gradient of normal $pm_2 = -1$, $m_2 = -\frac{1}{p}$.

Equation of normal at P, $y = -\frac{1}{p}x + c$, this passes through P,

$$ap^2 = -\frac{1}{p}2ap + c,$$

$c = ap^2 + 2a$ therefore

$$y = -\frac{1}{p}x + ap^2 + 2a$$

$$\boxed{py + x = ap^3 + 2ap}$$

Point of Intersection of the Tangents to $P(2ap, ap^2)$ and $Q(2aq, aq^2)$

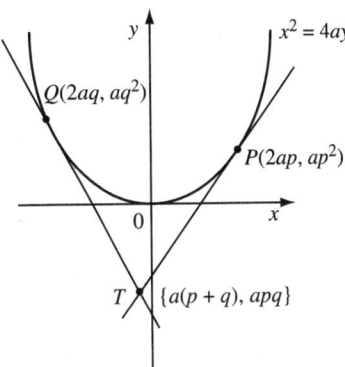

Fig. 6-I/38 Intersection of tangents at T

The equations of the tangents at $P(2ap, ap^2)$ and $Q(2aq, aq^2)$ are $y = px - ap^2$ and $y = qx - aq^2$ respectively.

Solving these equations simultaneously
$px - ap^2 = qx - aq^2$, $(p-q)x = a(p-q)(p+q)$

$\boxed{x = a(p+q)}$, substituting in one of the above equations for x, we have

$$y = p[a(p+q)] - ap^2$$
$$= ap^2 + apq - ap^2 = apq \quad \boxed{y = apq}$$

Thus, the tangents at $P(2ap, ap^2)$ and $Q(2aq, aq^2)$ meet at the point T with coordinates $T\{a(p+q), apq\}$.

Point of Intersection of the Normals to $P(2ap, ap^2)$ and $Q(2aq, aq^2)$

The equations of the normal at $P(2ap, ap^2)$ and $Q(2aq, aq^2)$ are $py + x = ap^3 + 2ap \ldots$ (1) and $qy + x = aq^3 + 2aq \ldots$ (2) respectively. Solving these equations simultaneously we have $py - qy = ap^3 + 2ap - aq^3 - 2aq$, by subtracting (2) from (1).

therefore $y(p - q) = a(p^3 - q^3) + 2a(p - q)$ hence $\boxed{y = a(p^2 + pq + q^2) + 2a}$ ...(3) where $p^3 - q^3 = (p - q)(p^2 + pq + q^2)$.

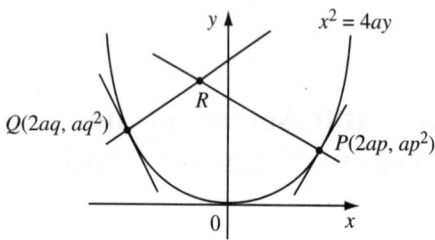

Fig. 6-I/39

Substituting (3) in (1), we have $pa(p^2 + pq + q^2) + 2ap + x = ap^3 + 2ap x = ap^3 + 2ap - ap^3 - ap^2q - apq^2 - 2ap$
$\boxed{x = -apq(p + q)}$

Therefore, the normals at $P(2ap, ap^2)$ and $Q(2aq, aq^2)$ meet at the point R with coordinates $R\left[-apq(p + q), a(p^2 + pq + q^2 + 2)\right]$

The Focal Chord

The equation of chord PQ is $2y = (p + q)x - 2apq$. When this chord passes through the focus $S(0, a)$ then it is called the focal chord.

Substituting $x = 0$ and $y = a$ in the equation of chord, we have $2a = -2apq$,

$\boxed{pq = -1}$. Hence, this is the condition for the chord PQ to pass through the focus.

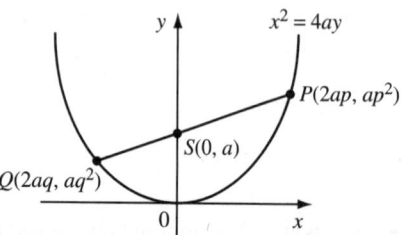

Fig. 6-I/40 The Focal Chord

The tangents at P and Q meet at $T\{a(p + q), apq\}$. If PQ is a focal chord, then $pq = -1$, and hence $T\{a(p + q), -a\}$. Since the equation of the directrix is $y = -a$, then T lies on the directrix of the parabola.

The tangents at the points $P(2ap, ap^2)$, $Q(2aq, aq^2)$ of the focal chord are at right angles since $pq = -1$ where p and q are the gradients of the tangents

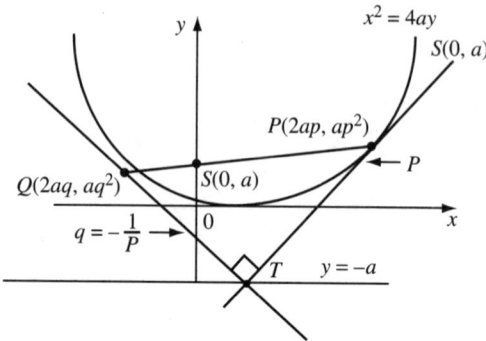

Fig. 6-I/41 The tangent at P and Q of the focal chord are at right angles.

Condition for a Straight Line $y = mx + c$ to Touch $x^2 = 4ay$

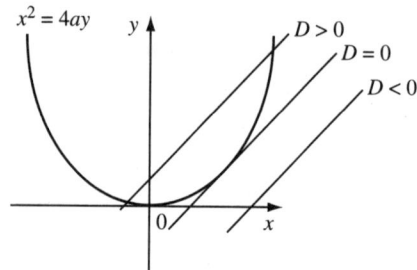

Fig. 6-I/42

What is the condition that the straight line $y = mx + c$ is tangent to the curve $x^2 = 4ay$?

Substituting $y = mx + c$ in $x^2 = 4ay$, we have
$$x^2 = 4a(mx + c)$$
$$x^2 - 4amx - 4ac = 0.$$

Examine the discriminant of this quadratic equation. If the discriminant, $D = b^2 - 4ac \geq 0$, the roots are real, $D = (-4am)^2 - 4(-4ac) \geq 0$. The straight line touches the curve when $D = 0$

$\boxed{D = 16a^2m^2 + 16ac = 0}$

If $D > 0$, the straight line intersects the parabola. If, however, the straight line neither intersects nor touches the curve, the discriminant, $D < 0$.

Therefore, for the straight line to touch the curve

$$D = 16a^2m^2 + 16ac = 0 \text{ or } 16a(am^2 + c) = 0.$$

Since a and m^2 are positive then $\boxed{c = -am^2}$ which is a negative quantity.

Hence the line $y = mx - am^2$ is a tangent to $x^2 = 4ay$ for all values of m.

Equation of the Chord of Contact from (x_1, y_1) to $x^2 = 4ay$

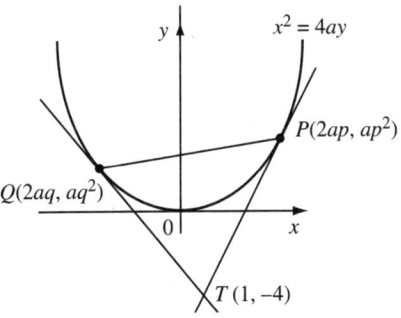

Fig. 6-I/43

Two tangents are drawn from the point $T(1, -4)$ to touch the parabola $x^2 = 4ay$ at P and Q.

The chord of contact is PQ, then the equation of the chord PQ is $2y - (p+q)x + 2apq = 0$.

The tangents at P and Q intersect in $T\{a(p+q), apq\}$ and since the coordinates of T are $(1, -4)$

$$1 = a(p+q),$$
$$-4 = apq$$
$$p + q = \frac{1}{a},$$
$$apq = -4,$$

the equations of PQ is $2y - \frac{1}{a}x + 2(-4) = 0$

or $2ay - x - 8a = 0$.

If $T(x_1, y_1)$ then $2ay - xx_1 + 2ay_1 = 0$ or
$$xx_1 = 2a(y + y_1).$$

Thus, the equation of the chord of contact of tangents drawn from (x_1, y_1) to the parabola $x^2 = 4ay$ is $\boxed{xx_1 = 2a(y + y_1)}$

WORKED EXAMPLE 22

Determine the equations of the parabolas with parametric equations

(i) $x = 5p, \quad y = 6p^2$
(ii) $x = 5t^2, \quad y = 3t$
(iii) $x = -3q, \quad y = 4q^2$
(iv) $x = -3p^2, \quad y = 5p$
(v) $x = -2t, \quad y = -3t^2$
(vi) $x = 3t^2, \quad y = 6t$.

Sketch the parabolas.

Solution 22

(i) $x = 5p, y = 6p^2$. Eliminating the parameter, p, between these two equations, we have

$$x = 5p, p = \frac{x}{5}; y = 6\left(\frac{x}{5}\right)^2$$

$$y = \frac{6}{25}x^2 \quad \text{or} \quad \boxed{6x^2 = 25y}$$

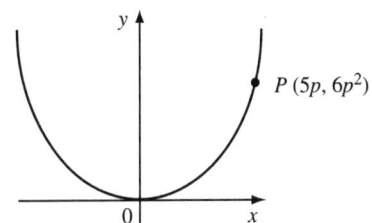

Fig. 6-I/44

(ii) $x = 5t^2, y = 3t$. Eliminating the parameter, t, between these two equations, we have

$$x = 5t^2 = 5\left(\frac{y}{3}\right)^2 = \frac{5y^2}{9} \quad \boxed{5y^2 = 9x}$$

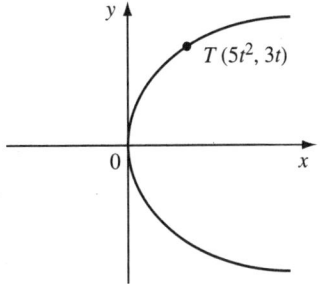

Fig. 6-I/45

(iii) $x = -3q$, $y = 4q^2$. Eliminating the parameter q, we have $y = 4q^2$

$$= 4\left(-\frac{x}{3}\right)^2$$

$$= \frac{4}{9}x^2$$

$\boxed{4x^2 = 9y}$

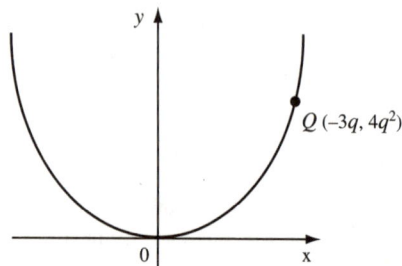

Fig. 6-I/46

(iv) $x = -3p^2$, $y = 5p$. Eliminating the parameter p, the locus is given as

$x = -3p^2$

$$= -3\left(\frac{y}{5}\right)^2 = -\frac{3y^2}{25}$$

or $\boxed{3y^2 = -25x}$

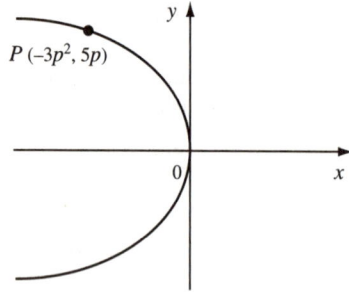

Fig. 6-I/47

(v) $x = -2t$, $y = -3t^2$. Eliminating t, the locus is given

$$y = -3\left(-\frac{x}{2}\right)^2 = -\frac{3x^2}{4}$$

or $\boxed{3x^2 = -4y}$

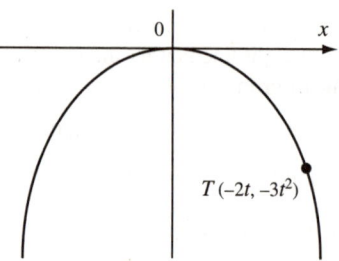

Fig. 6-I/48

(vi) $x = 3t^2$, $y = 6t$. Eliminating t, the locus is given by $x = 3t^2 = 3\left(\frac{y}{6}\right)^2$ or $36x = 3y^2$ $3y^2 = 36x$

or $\boxed{y^2 = 12x}$

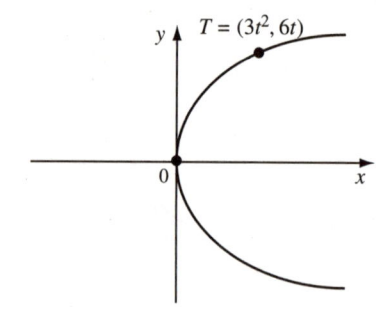

Fig. 6-I/49

Properties of the Parabola

A parabola is symmetrical about the y-axis has its vertex at $V(0, 0)$ and has the standard equation $\boxed{x^2 = 4ay}$

if and only if its focus has the coordinates $(0, a)$ and its directrix has an equation $y = -a$.

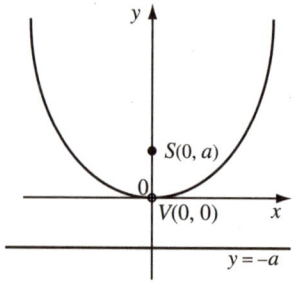

Fig. 6-I/50

A parabola is symmetrical about the x-axis has its vertex at $V(0, 0)$ and has the standard equation $\boxed{y^2 = 4ax}$

if and only if its focus has the coordinates $S(a, 0)$ and its directrix has an equation $x = -a$.

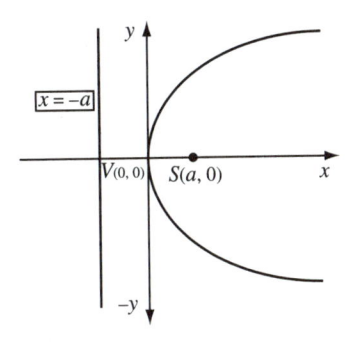

Fig. 6-I/51

The point (x, y) moves in such a way that its distance from the point $(0, 2)$ is equal to its distance form the line $y = -2$. The equation of the parabola generated is

$$\boxed{x^2 = 8y} \quad \ldots(1)$$

The point (x, y) moves in such a way that its distance from the point $(1, 0)$ is equal to its distance from the line $x = -1$. The equation of the parabola generated is

$$\boxed{y^2 = 4x} \quad \ldots(2)$$

The point (x, y) moves in such a way that its distance from the point $(0, -3)$ is equal to its distance from the line $y = 3$. The equation of the parabola generated is

$$\boxed{x^2 = -12y} \quad \ldots(3)$$

The point (x, y) moves in such a way that its distance from the point $(-4, 0)$ is equal to its distance from the line $x = 4$. The equation of the parabola generated is

$$\boxed{y^2 = -16x} \quad \ldots(4)$$

Sketch the four parabolas given by the equations (1), (2), (3) and (4), on the same graph and hatch the areas of the loops. The angle between two curves at A is the angle between the two tangents at A.

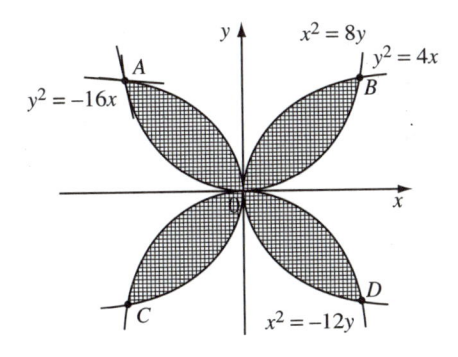

Fig. 6-I/52

Exercises 3

1. Show that the following points lie on the parabolas for all values of the parameters.

   (i) $P(5p, 6p^2)$, $6x^2 = 25y$

   (ii) $T(5t^2, 3t)$, $5y^2 = 9x$

   (iii) $Q(-3q, 4q^2)$, $4x^2 = 9y$

   (iv) $P(-3p^2, 5p)$, $3y^2 = -25x$

   (v) $T(-2t, -3t^2)$, $3x^2 = -4y$

   (vi) $T(3t^2, 6t)$, $y^2 = 12x$.

2. Show that the line $y = mx - 5m^2$ is a tangent to $x^2 = 20y$.

3. Show that the line $y = mx - 7m^2$ is a tangent to $x^2 = 28y$.

4. Show that the line $y = mx - 3m^2$ is a tangent to $x^2 = 12y$.

5. The tangents at $P(2ap, ap^2)$ and $Q(2aq, aq^2)$ on the parabola $x^2 = 4ay$ intersect at T. If M is the mid-point of the chord PQ, show that TM is parallel to the y-axis of the parabola.

6. If T is the point $(2at, at^2)$ on the parabola $x^2 = 4ay$. Prove that the equation of the normal at T is $x + ty = at^3 + 2at$. Determine the equation of the tangent at T.

7. Show that the normal at $P(ap^2, 2ap)$ has a gradient $-p$ and determine its equation.

8. The equation of a parabola is $y^2 = 4ax$. Find the coordinates of the focus and the equation of the directrix.

9. The equation of a parabola is $y^2 = -4ax$. Find the coordinates of the focus and the equation of the directrix.

10. The equation of a parabola is $x^2 = 4ay$. Find the coordinates of the focus and the equation of the directrix.

11. The equation of a parabola is $x^2 = -4ay$. Find the coordinates of the focus and the equation of the directrix.

12. Find the equations of the parabolas:

    (i) focus $(-2, 0)$, directrix $x = 2$,

    (ii) focus $(3, 0)$, directrix $x = -3$,

(iii) focus (0, 1), directrix $y = -1$,

(iv) focus (0, −4), directrix $y = 4$.

13. Find, in terms of a, m, the value of c which makes the line $y = mx + c$ a tangent to the parabola $x^2 = 4ay$ hence obtain the coordinates of the point of contact.

14. Find the values of m such that the line $y = mx + 1$ is a tangent to the parabola $(y-1)^2 = 4(x-2)$, hence obtain the coordinates of the points of contacts P and Q. Sketch these tangents on the parabola. Show that the length PQ is 2 units.

15. Determine the equation of the tangent to the parabola $x^2 = 4ay$ at (x_1, y_1).

16. Find the equation of the chord joining the points $T(2at, at^2)$, $P(2ap, ap^2)$ on the parabola $x^2 = 4ay$.

17. Find the coordinates of the point of intersection, T of the normals at the points $P(2ap, -ap^2)$ and $Q(2aq, -aq^2)$ of the parabola $x^2 = -4ay$. Determine the locus of T if $pq = -1$.

18. Find the equation of the chord joining the points $P(ap^2, 2ap)$ and $Q(aq^2, 2aq)$ of the parabola hence show that $pq = -1$ when the chord passes through the focus.

19. Find the coordinates of the point of intersection of the tangents at the points $P\left(-ap^2, 2ap\right)$ and $Q\left(-aq^2, 2aq\right)$ of the parabola $y^2 = -4ax$.

20. Find the equations of the tangents to the parabola $x^2 = 4ay$ from the point $(3a, -5a)$.

21. Find the focus and directrix of the parabola $x^2 = 4(y - 4)$ and give the length of its latus rectum.

22. Find the latus rectum for the following parabolas:

    (i) $x^2 = 8y$

    (ii) $y^2 = 4x$

    (iii) $(x-1)^2 = 8(y-3)$

    (iv) $(y+2)^2 = 4(x-1)$.

23. Find the equation of the tangent at the vertex of $x^2 = 8y - 8$.

24. Verify that the parametric equations of the parabola $(2y - 1)^2 = 3(x - 1)$ are: $x = 3t^2 + 1$ and $y = \frac{3}{2}t + \frac{1}{2}$.

25. Determine the focus and directrix of the parabola $y = x^2 + 2x + 3$.

26. Find the equation of the parabola whose focus is (0, 3) and directrix $y = -3$.

27. Determine the focus and directrix of the parabola $x = y^2 + 2y - 4$.

28. Find the point of intersection of the normals at the points $P(2p^2, 4p)$ and $Q(2q^2, 4q)$ of the parabola $y^2 = 8x$.

29. Determine the equations of the parabolas given that

    (i) the directrix $x = -2$ and the focus $S(2, 0)$

    (ii) the directrix $y = -3$ and the focus $S(0, 3)$

    (iii) the directrix $x = 5$ and the vertex $V(3, 2)$

    (iv) the directrix $y = 5$ and the focus $S(0, -5)$

    (v) the directrix $x = 0$ and the focus $S(4, 4)$

    (vi) the vertex $V(-3, -4)$ and $S(-3, -6)$.

30. For the following equations of parabolas, find the corresponding directrices and foci. Sketch the curves.

    (i) $y^2 = 4x$

    (ii) $y^2 = -5x$

    (iii) $x^2 = -4y$

    (iv) $x^2 = 9y$

    (v) $5y^2 = 2x$

    (vi) $7x^2 = -3y$.

31. Find the point of intersection of the normals at the points $P(ap^2, 2ap)$ and $Q(aq^2, 2aq)$ of the parabola $y^2 = 4ax$.

32. Find the equation of the parabola whose focus is (0, 3) and directrix $y = -3$.

33. Find the points on the parabola $y^2 = 4x$ where

    (i) the tangent and

    (ii) the normal are parallel to the line $2x + y = 1$.

4

To Derive the Equation of the Ellipse

$$\boxed{\frac{x^2}{a^2} + \frac{y^2}{b^2} = 1} \quad (a > b > 0)$$

Starting with the definition, given a fixed point, S, the focus, and a fixed line, the <u>directrix</u>, $\frac{SP}{PN} = e (e < 1)$

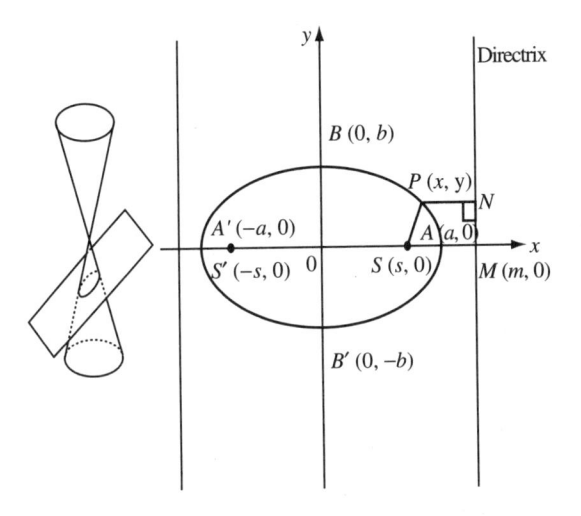

Fig.6-1/53

where e is the <u>eccentricity</u> of the ellipse. <u>Note</u> that the eccentricity of the parabola is $e = 1$, and that of the circle is $e = 0$.

Let A be $(a, 0)$, $A'(-a, 0)$, $B(0, b)$, $B'(0, -b)$ from the equation of the ellipse $\frac{x^2}{a^2} + \frac{y^2}{b^2} = 1$, if $x = 0$, $y = \pm b$; and if $y = 0$, $x = \pm a$.

A and A' are called the vertices.

$\frac{SA}{AM} = e, a - s = SA = (m - a)e = AMe$

$a - s = e(m - a)$ ...(1)

$\frac{SA'}{A'M} = e, SA' = a + s = eA'M = e(m + a)$

$a + s = e(m + a)$ ...(2)

Adding equations (1) and (2)

$2a = 2em$

$m = \frac{a}{e}$

Subtracting (1) from (2)

$2s = 2ae$

$s = ae$.

The coordinates of S are $(ae, 0)$ and the equation of the directrix is $x = \frac{a}{e}$.

Let $P(x, y)$, $\frac{SP}{PN} = e$

$SP^2 = e^2 PN^2$

but $SP^2 = (x - ae)^2 + y^2$ and $PN = \frac{a}{e} - x$.

Therefore, $(x - ae)^2 + y^2 = e^2 \left(\frac{a}{e} - x\right)^2$ expanding

$x^2 - 2aex + a^2e^2 + y^2 = \frac{e^2 a^2}{e^2} - e^2 \frac{2a}{e} x + e^2 x^2$

$x^2 \left(1 - e^2\right) + y^2 = a^2 \left(1 - e^2\right)$

dividing each term by $a^2(1 - e^2)$, we have

$\frac{x^2}{a^2} + \frac{y^2}{a^2(1 - e^2)} = 1$

therefore $\boxed{\frac{x^2}{a^2} + \frac{y^2}{b^2} = 1}$

where $b^2 = a^2 \left(1 - e^2\right)$.

If x is replaced by $-x$, we have found the other focus, $S'(-ae, 0)$ and the other directrix, $x = \frac{a}{e}$.

There are two foci, $S(ae, 0)$ and $S'(-ae, 0)$ and two directrices, $x = \frac{a}{e}$ and $x = \frac{a}{e}$. The ellipse has two axes the <u>minor axis</u> and the major axis in this case, the minor axis is $BB'(2b)$ and the major axis is $AA'(2a)$. If $b > a$, however, the minor axis is the AA' and the major axis is the BB'. The axes of <u>symmetry</u> meet at the centre of the ellipse. Any chord passing through the centre is called a <u>diameter</u>. The circle on AA' as diameter is auxiliary circle of the ellipse.

WORKED EXAMPLE 23

If the eccentricity of an ellipse is $\frac{2}{3}$ and $a = 5$, determine the value of b and hence write down the equation of the curve.

Solution 23

$$b^2 = a^2(1-e^2) = 5^2\left[1 - \left(\frac{2}{3}\right)^2\right] = 25\left(1 - \frac{4}{9}\right)$$

$$= 25 \times \frac{5}{9}$$

$$b^2 = \frac{125}{9}, \quad b = \pm\frac{5}{3}\sqrt{5}.$$

$$\frac{x^2}{a^2} + \frac{y^2}{b^2} = 1,$$

$$\frac{x^2}{5^2} + \frac{y^2}{\frac{125}{9}} = 1.$$

$$\boxed{5x^2 + 9y^2 = 125}$$

WORKED EXAMPLE 24

Find the coordinates of the foci of the ellipse $5x^2 + 9y^2 = 125$, and the equation of the directrices.

Solution 24

The foci are $S(ae, 0)$ and $S'(-ae, 0)$

$$a = 5, e = \frac{2}{3}$$

$$S\left(\frac{10}{3}\right) \text{ and } S'\left(-\frac{10}{3}, 0\right).$$

The directrices are given

$$x = \frac{a}{e} \text{ and } x = -\frac{a}{e}, \text{ therefore}$$

$$x = \frac{5}{\frac{2}{3}} = \frac{15}{2} \text{ and } x = -\frac{15}{2}$$

Eccentric Angles

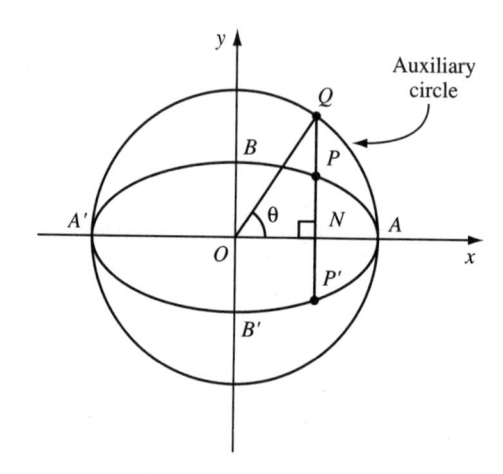

Fig.6-I/54

The two axes of symmetry $A'OA$, $B'OB$ are called the principal axes of the ellipse, their point of intersection O is called the <u>centre</u> and any chord through the centre is called a <u>diameter</u>. A and A' are called the <u>vertices</u>, AA', length $2a$, is called the <u>major axis</u> $B'B$, length $2b$, is called the <u>minor axis</u> PNP' is called a double <u>ordinate</u> where PN is the ordinate. The circle on AA' as diameter is called the auxiliary circle. The ordinate PN is produced to meet the auxiliary circle at Q, the angle $\angle QON = \theta = $ eccentric angle of P.

$$NQ = a\sin\theta,$$

$$NP = \frac{b}{a}.NQ = b\sin\theta.$$

Therefore, Q is the point $(a\cos\theta, a\sin\theta)$ and P is the point $(a\cos\theta, b\sin\theta)$.

Hence any point $P(x, y)$ on the ellipse $\frac{x^2}{a^2} + \frac{y^2}{b^2} = 1$ is given in the parametric equations $x = a\cos\theta$

$$y = b\sin\theta$$

$(a > b > 0)$.

To Derive the Equation of the Ellipse — 35

The gradient of the tangent at any point is $\frac{dy}{dx} = -\frac{b^2}{a^2}\frac{x}{y}$, the gradient at $P(x_1, y_1)$ is $\frac{dy}{dx} = -\frac{b^2}{a^2}\frac{x_1}{y_1}$.

The equation of the tangent $y = \frac{dy}{dx} = -\frac{b^2}{a^2}\frac{x_1}{y_1} \cdot x + c$. this passes through the point $P(x_1, y_1)$,

$$y_1 = -\frac{b^2}{a^2}\frac{x_1}{y_1}x_1 + c,$$

$$c = \frac{y_1^2 a^2 + b^2 x_1^2}{a^2 y_1},$$

$$y = -\frac{b^2}{a^2}\frac{x_1}{y_1}x + \frac{y_1^2 a^2 + b^2 x_1^2}{a^2 y_1}$$

$$ya^2 y_1 + b^2 x_1 x = y_1^2 a^2 + b^2 x_1^2$$

$$\frac{yy_1 a^2}{a^2 b^2} + \frac{xx_1 b^2}{a^2 b^2} = \frac{y_1^2 a^2}{a^2 b^2} + \frac{b^2 x_1^2}{a^2 b^2}$$

$$\frac{yy_1}{b^2} + \frac{xx_1}{a^2} = \frac{y_1^2}{b^2} + \frac{x_1^2}{a^2} = 1 \text{ therefore}$$

$$\boxed{\frac{xx_1}{a^2} + \frac{yy_1}{b^2} = 1}$$

WORKED EXAMPLE 25

If α is the angle between the ellipse and the circle as shown in Fig. 6-I/53. Determine the area of the ellipse.

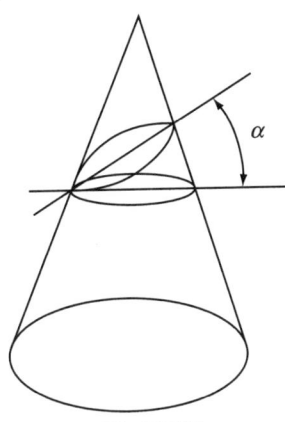

Fig.6-I/55

Solution 25

$$\cos\alpha = \frac{\text{area of the ellipse}}{\text{area of the circle}}$$

but $\cos\alpha = \frac{b}{a} = \frac{\text{area of ellipse}}{\text{area of circle}}$

$$= \frac{\text{area of the ellipse}}{\pi a^2}$$

area of the ellipse $= \pi a^2 \frac{b}{a} = \pi ab$.

Determine the Equation of the Tangent at $P(x_1, y_1)$ to the Ellipse $\frac{x^2}{a^2} + \frac{y^2}{b^2} = 1$

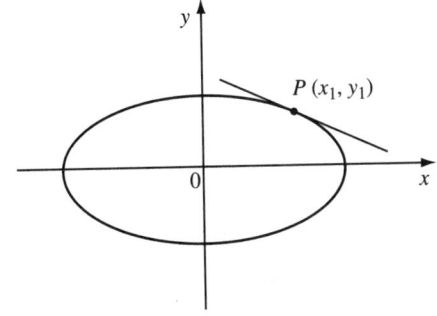

Fig.6-I/56

Determine the Tangent to the Ellipse $\frac{x^2}{a^2} + \frac{y^2}{b^2} = 1$ at the Point $P(a\cos\theta, b\sin\theta)$

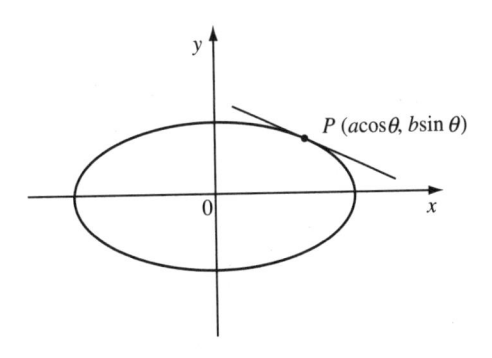

Fig.6-I/57

$x = a\cos\theta \qquad \frac{dx}{d\theta} = -a\sin\theta$

$y = b\sin\theta \qquad \frac{dy}{d\theta} = b\cos\theta$

$\frac{dy}{dx} = \frac{\frac{dy}{d\theta}}{\frac{dx}{d\theta}} = \frac{b\cos\theta}{-a\sin\theta} = -\frac{b}{a}\cot\theta$ the gradient of the tangent at P.

Alternatively

Differentiating $\frac{x^2}{a^2} + \frac{y^2}{b^2} = 1$ implicitly $\frac{2x}{a^2} + \frac{2y}{b^2}\frac{dy}{dx} = 0$ or $\frac{dy}{dx} = -\frac{b^2}{a^2}\frac{x}{y}$ the gradient of the tangent at any point, at P, the gradient is

$\frac{dy}{dx} = -\frac{b^2}{a^2} \cdot \frac{a\cos\theta}{b\sin\theta} = -\frac{b}{a}\cot\theta$ $\boxed{\frac{dy}{dx} = -\frac{b}{a}\cot\theta}$

The equation of the tangent is $y = mx + c$, $y = -\frac{b}{a}\cot\theta \, x + c$, this passes through the point P, $b\sin\theta = -\frac{b}{a}\cot\theta(a\cos\theta) + c$

$c = b\sin\theta + b\frac{\cos^2\theta}{\sin\theta} = \frac{b(\sin^2\theta + \cos^2\theta)}{\sin\theta}$

$c = \frac{b}{\sin\theta}, \quad y = -\frac{b}{a}\cot\theta x + \frac{b}{\sin\theta}$

$\boxed{ay\sin\theta + bx\cos\theta = ab}$

Determine the Normal to the Ellipse $\frac{x^2}{a^2} + \frac{y^2}{b^2} = 1$ at the Point $P(a\cos\theta, b\sin\theta)$

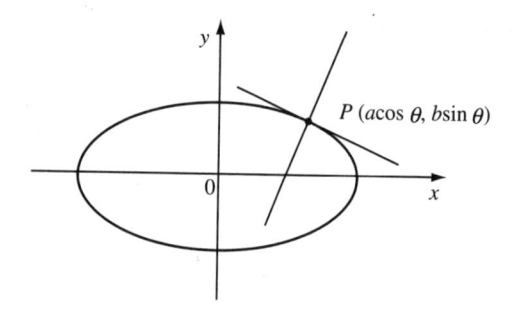

Fig.6-I/58

The gradient of the tangent at any point to the ellipse is $\frac{dy}{dx} = -\frac{b^2}{a^2}\frac{x}{y}$, the normal at any point is given from $m_1 m_2 = -1$ as $\frac{dy}{dx} = \frac{a^2}{b^2}\frac{y}{x}$. The gradient of the normal at P is $\frac{dy}{dx} = \frac{a^2}{b^2} \cdot \frac{b\sin\theta}{a\cos\theta} = \frac{a}{b}\tan\theta$

$\boxed{\frac{dy}{dx} = \frac{a}{b}\tan\theta}$

The equation of the normal is $y = \frac{a}{b}\tan\theta x + c$ this passes through P,

$b\sin\theta = \frac{a}{b}\tan\theta \cdot a\cos\theta + c$

$c = \frac{b^2\sin\theta - a^2\sin\theta}{b} = \frac{\sin\theta}{b}\left(b^2 - a^2\right)$

$y = \frac{a}{b}x\tan\theta + \frac{\sin\theta}{b}\left(b^2 - a^2\right)$

$yb = ax\tan\theta + \sin\theta\left(b^2 - a^2\right)$

$\boxed{yb\cos\theta = ax\sin\theta + \sin\theta\cos\theta\left(b^2 - a^2\right)}$

Determine the equation of the Chord of the Ellipse $\frac{x^2}{a^2} + \frac{y^2}{b^2} = 1$ between the Points $P(a\cos\theta, b\sin\theta)$ and $Q(a\cos\phi, b\sin\phi)$

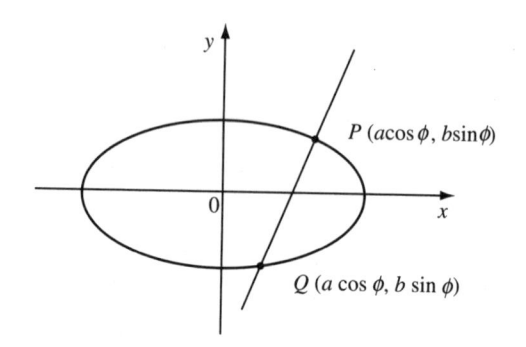

Fig.6-I/59

Where θ and ϕ are the eccentric angles. The gradient of the chord PQ is given

$m = \frac{b\sin\theta - b\sin\phi}{a\cos\theta - a\cos\phi} = \frac{b(\sin\theta - \sin\phi)}{a(\cos\theta - \cos\phi)}$

$= \frac{2\cos\frac{\theta+\phi}{2}\sin\frac{\theta-\phi}{2}}{-2\sin\frac{\theta+\phi}{2}\sin\frac{\theta-\phi}{2}} \cdot \frac{b}{a} = -\frac{b}{a} \cdot \frac{\cos\frac{\theta+\phi}{2}}{\sin\frac{\theta+\phi}{2}}$

The equation of the chord is $y = mx + c$

$y = -\frac{b\cos\frac{\theta+\phi}{2}}{a\sin\frac{\theta+\phi}{2}}x + c,$

this passes through P or Q, through P, b

To Derive the Equation of the Ellipse — 37

$$\sin\theta = -\frac{b\cos\frac{\theta+\phi}{2}}{a\sin\frac{\theta+\phi}{2}}a\cos\theta + c$$

$$c = \frac{1}{\sin\frac{\theta+\phi}{2}}\left(b\sin\theta\sin\frac{\theta+\phi}{2} + b\cos\frac{\theta+\phi}{2}\cos\theta\right),$$

therefore

$$y = -\frac{b\cos\frac{\theta+\phi}{2}}{a\sin\frac{\theta+\phi}{2}}x$$

$$+ \left(b\sin\theta\sin\frac{\theta+\phi}{2} + b\cos\frac{\theta+\phi}{2}\cos\theta\right)\frac{a}{a\sin\frac{\theta+\phi}{2}}.$$

Therefore the equation of the chord is

$$\boxed{bx\cos\frac{\theta+\phi}{2} + ay\sin\frac{\theta+\phi}{2} - ab\cos\frac{\theta-\phi}{2} = 0}$$

where $ab\cos\frac{\theta+\phi}{2}\cos\theta + ab\sin\frac{\theta+\phi}{2}\sin\theta$

$$= ab\cos\left(\frac{\theta+\phi}{2} - \theta\right)$$

$$= ab\cos\frac{\phi-\theta}{2}.$$

Observe that if $\phi = \theta$, the equation of the chord approaches the equation of the tangent at θ as $\phi \to \theta$.

WORKED EXAMPLE 26

Find the equation of the normal at the point $P(3\cos\phi, \sqrt{2}\sin\phi)$ to the ellipse. If this normal passes through the focus $S(\sqrt{7}, 0)$, show that $\cos\phi = \frac{3}{\sqrt{7}}$.

Solution 26

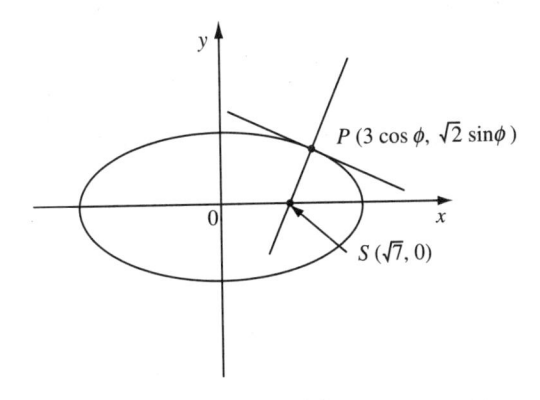

Fig.6-I/60

$x = 3\cos\phi, \frac{dx}{d\phi} = -3\sin\phi$

$a = 3, ae = \sqrt{7}, e = \frac{\sqrt{7}}{3}$

$y = \sqrt{2}\sin\phi, \frac{dy}{d\phi} = \sqrt{2}\cos\phi$

$b = \sqrt{2}$

$\frac{dy}{dx} = -\frac{\sqrt{2}}{3}\cot\phi$, the gradient of the normal is $\frac{dy}{dx} = \frac{3}{\sqrt{2}}\tan\phi$, the equation of the normal $y = \frac{3}{\sqrt{2}}(\tan\phi)x + c$, $\sqrt{2}\sin\phi = \frac{3}{\sqrt{2}}\tan\phi(3\cos\phi) + c$

$c = \sqrt{2}\sin\phi - \frac{9}{\sqrt{2}}\sin\phi$

$y = \frac{3}{\sqrt{2}}(\tan\phi)x + \left(\sqrt{2} - \frac{9}{\sqrt{2}}\right)\sin\phi$

$\sqrt{2}y = 3(\tan\phi)x + (2-9)\sin\phi$

$x = \sqrt{7}, y = 0; 0 = 3(\tan\phi)\sqrt{7} - 7\sin\phi$

$3\sqrt{7}\tan\phi = 7\sin\phi$

$3\sqrt{7} = 7\cos\phi$,

$\cos\phi = \frac{3\sqrt{7}}{7} \times \frac{\sqrt{7}}{\sqrt{7}} = \frac{3}{\sqrt{7}}.$

Exercises 4

1. $P(a\cos\theta, b\sin\theta)$ is any point on the ellipse $b^2x^2 + a^2y^2 = a^2b^2$.

   (i) Verify that this point lies on the ellipse.

   (ii) Determine the equation of the tangent at this point.

   (iii) Determine the equation of the normal at this point.

   (iv) Determine the equation of the line which is parallel to the tangent of (ii) and passes through the origin.

   (v) Determine the equation of the line that is parallel to the normal of (iii) and passes through the origin.

2. Find the equations of the tangents from (x_1, y_1) to the ellipse $b^2x^2 + a^2y^2 = a^2b^2$.

3. Determine the coordinates of the point of intersection of the tangents at the points $P(3\cos\alpha, 4\sin\alpha)$ and $Q(3\cos\beta, 4\sin\beta)$ of the ellipse $16x^2 + 9y^2 = 144$.

4. Prove that the line $y = mx + 5$ is a tangent to the ellipse $9x^2 + 16y^2 = 144$ provided that $m = \pm 1$.

5. Find the equations of the tangents from P(6, 8) to the ellipse $9x^2 + 25y^2 = 225$.

6. The equations of the ellipses are given:

   (i) $\dfrac{x^2}{4} + \dfrac{y^2}{3} = 1$

   (ii) $x^2 + \dfrac{y^2}{2} = 1$

   (iii) $\dfrac{x^2}{3^2} + \dfrac{y^2}{4^2} = 1$

   (iv) $\dfrac{x^2}{5^2} + \dfrac{y^2}{3^2} = 1$

   (v) $4x^2 + 8y^2 = 32$.

   (a) Find the coordinates of the foci.
   (b) Find the eccentricities of the curves.
   (c) Write down the equation of the directrices.

7. Find the foci and directrices of the ellipses:

   (i) $8x^2 + 12y^2 = 96$

   (ii) $2x^2 + 16y^2 = 32$

   (iii) $4x^2 + y^2 = 4$.

8. Find the equation of the tangent to the following curves:

   (i) $5x^2 + y^2 = 25$ at $(2, \sqrt{5})$

   (ii) $x^2 + 4y^2 = 8$ at $(-2, 1)$

9. A variable normal to the ellipse meets the axes at P and Q. Find the locus of the mid-point of PQ.

10. Prove that the perpendicular tangents to the ellipse $b^2x^2 + a^2y^2 = a^2b^2$ meet on the director circle of the ellipse. What is the difference between an auxiliary and director circles.

11. Find the equation of the line with gradient m passing through the focus $(-ae, 0)$ of the ellipse $b^2x^2 + a^2y^2 = a^2b^2$.

12. Find the equation of the tangent at the point (x_1, y_1) on the ellipse $b^2x^2 + a^2y^2 = a^2b^2$.

13. A straight line $y = mx + c$ intersects the ellipse $b^2x^2 + a^2y^2 = a^2b^2$ at the points $P(a\cos\theta, b\sin\theta)$ and $Q(a\cos\phi, b\sin\phi)$. Show that $m = -\dfrac{b}{a}\cot\dfrac{\theta+\phi}{2}$. If $M(x, y)$ are the coordinates of the mid-point of P and Q find an expression for $x^2 + y^2$ in terms of a, b, θ and ϕ.

14. The line $2y = x + 6$ and the ellipse $25x^2 + 36y^2 = 900$ intersect, show that the x coordinates of the points of intersection are given by the solutions of the quadratic equation $17x^2 + 54x - 288 = 0$.

15. Find the equations of the tangent and normal at the given point:

   (i) $8x^2 + 12.5y^2 = 52$; $(2, -2)$

   (ii) $9x^2 + 2y^2 = 54$; $(-2, 3)$

   (iii) $9x^2 + 4y^2 = 100$; $(2, 4)$

   (iv) $3x^2 + y^2 = 12$; $(-1, 3)$

   (v) $9x^2 + 2y^2 = 9$; $\left(\dfrac{1}{3}, 2\right)$.

16. Prove that $x + y = 2$ is a normal to the ellipse $3x^2 + y^2 = 12$. Find the coordinates of the points of the intersections of this normal with the ellipse and hence find the equation of the other normal. Sketch the graph.

17. Find the coordinates of the points of contact of tangents to $b^2x^2 + a^2y^2 = a^2b^2$ parallel to $y = -x$ ($a = 3$ and $b = 4$).

18. Find the equation of the normals to $b^2x^2 + a^2y^2 = a^2b^2$ which are parallel to $3x + y + 1 = 0$ ($a = 9$ and $b = 4$).

19. Determine the equation of the chord of the ellipse $9x^2 + 4y^2 = 36$ through the points $P(2\cos\alpha, 3\sin\alpha)$ and $Q(2\cos\beta, 3\sin\beta)$.

   Ans. $3x\cos\dfrac{\alpha+\beta}{2} + 2y\sin\dfrac{\alpha+\beta}{2} = 6\cos\dfrac{\alpha-\beta}{2}$.

20. Determine the equations of the tangents from the point $T(-1, 3)$ to the ellipse $\dfrac{x^2}{3^2} + \dfrac{y^2}{2^2} = 1$ and find the points of contacts.

5

The Hyperbola

Definition Derivation of the Equation of the Hyperbola

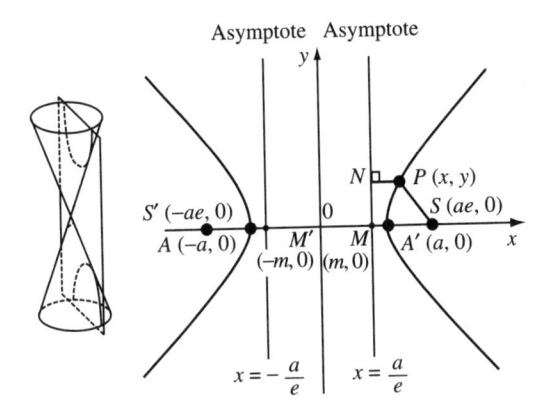

Fig. 6-I/61

DEFINITION

Let e be the eccentricity of the hyperbola, $e = \frac{SP}{PN}$ where SP is the distance of the locus from a fixed point $(s, 0)$ the focus and PN is the distance of the locus from a fixed line, the directrix which intersects the x-axis at $M(m, 0)$. If A is one vertex, $A(a, 0)$ then $SP = e\,PN$ or $SA = e\,AM$, $SA' = e\,AM'$

$SA = e\,AM$, $\quad s - a = e(a - m)$ ...(1)

$SA' = e\,AM'$, $\quad s + a = e(a + m)$ ...(2)

Adding (1) and (2) $\quad 2s = 2ea$

$$\boxed{s = ae}$$

Subtracting (1) and (2) $2a = 2em$

$$\boxed{m = \frac{a}{e}}$$

Therefore the coordinates of the focus are $S(ae, 0)$ and the equation of the directrix is $x = \frac{a}{e}$.

The foci are $(ae, 0)$ and $(-ae, 0)$ and the directrices are

$x = \frac{a}{e}$ and $x = -\frac{a}{e}$. $\frac{SP}{PN} = e$, $\frac{(SP)^2}{(PN)^2} = e^2$,

$(SP)^2 = y^2 + (ae - x)^2$ and $(PN)^2 = \left(x - \frac{a}{e}\right)^2$

therefore

$$y^2 + (ae - x)^2 = e^2 \left(x - \frac{a}{e}\right)^2$$

$$y^2 + a^2 e^2 - 2aex + x^2 = e^2 x^2 - 2axe + a^2$$

$$y^2 - e^2 x^2 + x^2 = -a^2 e^2 + a^2$$

$$y^2 - x^2(e^2 - 1) = a^2(1 - e^2)$$

$$\frac{y^2}{a^2(1 - e^2)} - \frac{x^2(e^2 - 1)}{a^2(1 - e^2)} = 1$$

$$\frac{x^2}{a^2} + \frac{y^2}{a^2(e^2 - 1)} = 1$$

let $b^2 = a^2(e^2 - 1)$

$$\boxed{\frac{x^2}{a^2} - \frac{y^2}{b^2} = 1}$$

The Parametric Equations of the Hyperbola

Trigonometric

$x = a\sec\theta$

$y = b\tan\theta$

Verify

$$\frac{x^2}{a^2} - \frac{y^2}{b^2} = 1$$

$$\frac{a^2 \sec^2 \theta}{a^2} - \frac{b^2 \tan^2 \theta}{b^2} = 1$$

$$\sec^2 \theta - \tan^2 \theta = 1$$

$$\boxed{1 + \tan^2 \theta = \sec^2 \theta}.$$

Hyperbolic

$$x = a\cosh \theta \qquad y = b\sinh \theta$$

$$\frac{a^2 \cosh^2 \theta}{a^2} - \frac{b^2 \sinh^2 \theta}{b^2} = 1$$

$$\boxed{\cosh^2 \theta - \sinh^2 \theta = 1}$$

The Equation of the Tangent at the Point $P\ (a\sec \theta, b\tan \theta)$

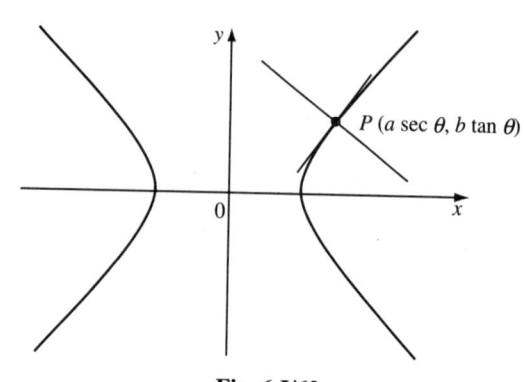

Fig. 6-I/62

$$x = a\sec \theta, \quad \frac{dx}{d\theta} = a\sec \theta \tan \theta$$

$$y = b\tan \theta, \quad \frac{dy}{d\theta} = b\sec^2 \theta$$

$$\frac{\frac{dy}{d\theta}}{\frac{dx}{d\theta}} = \frac{dy}{dx} = \frac{b\sec^2 \theta}{a\sec \theta \tan \theta} = \frac{b}{a}\csc \theta$$

the gradient of the tangent at $P(a\sec \theta, b\tan \theta)$. The equation of the tangent at $P(a\sec \theta, b\tan \theta)$ is $y = \frac{b}{a}x \csc \theta + c$, this passes through the point P,

$$b\tan \theta = \frac{b}{a} \csc \theta a\sec \theta + c$$

$$c = b\tan \theta - b\csc \theta \sec \theta$$

$$y = \frac{b}{a}(\csc \theta)x + b(\tan \theta - \csc \theta \sec \theta) \text{ multiply each term by } (a\sin \theta)$$

$$ay\sin \theta = bx + ab\left(\frac{\sin^2 \theta}{\cos \theta} - \sec \theta\right)$$

$$ay\sin \theta = bx + ab\left(\frac{\sin^2 \theta - 1}{\cos \theta}\right) = bx + ab(-\cos \theta)$$

$$\boxed{bx - ay\sin \theta - ab\cos \theta = 0}$$

The Equation of the Normal at the Point $P\ (a\sec \theta, b\tan \theta)$

The gradient of the normal is $-\frac{a}{b}\sin \theta$ and the equation of the normal $y = -\frac{a}{b}(\sin \theta)x + c$, this passes through P.

$$b\tan \theta = -\frac{a}{b}(\sin \theta)a\sec \theta + c$$

$$\Rightarrow c = \left(b\tan \theta + \frac{a^2}{b} \sin \theta \sec \theta\right)$$

$$y = -\frac{a}{b}(\sin \theta)x + b\tan \theta + \frac{a^2}{b} \sin \theta \sec \theta$$

multiplying each term by b

$$yb = -a(\sin \theta)x + b^2 \tan \theta + a^2 \sin \theta \sec \theta$$

$$\boxed{ax\sin \theta + yb - (a^2 + b^2)\tan \theta = 0}$$

The Equation of the Tangent at the Point $P(x_1, y_1)$ of the Hyperbola

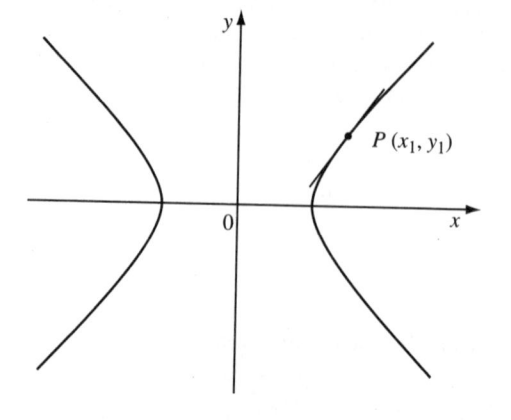

Fig. 6-I/63

The Hyperbola — 41

$\dfrac{x^2}{a^2} - \dfrac{y^2}{b^2} = 1$, differentiating with respect to x

$\dfrac{2x}{a^2} - \dfrac{2y}{b^2}\dfrac{dy}{dx} = 0,$

$\dfrac{dy}{dx} = \dfrac{2x}{a^2}\dfrac{b^2}{2y} = \dfrac{x}{y}\dfrac{b^2}{a^2}$, at P

$\dfrac{dy}{dx} = \dfrac{x_1}{y_1}\dfrac{b^2}{a^2}$, the gradient of the tangent.

$y = \dfrac{x_1}{y_1}\dfrac{b^2}{a^2}x + c,$

$c = y_1 - \dfrac{x_1}{y_1}\dfrac{b^2}{a^2}x_1 = \dfrac{a^2 y_1^2 - x_1^2 b^2}{a^2 y_1}$

$y = \dfrac{x_1}{y_1}\dfrac{b^2}{a^2}x + \dfrac{a^2 y_1^2 - x_1^2 b^2}{a^2 y_1}$

multiplying each term by $\dfrac{y_1 a^2}{b^2}$

$yy_1\dfrac{a^2}{b^2} = x_1 x + \dfrac{a^2 y_1^2 - b^2 x_1^2}{b^2}$

and dividing each term by a^2

$\dfrac{xx_1}{a^2} - \dfrac{yy_1}{b^2} = -\dfrac{a^2 y_1^2 - b^2 x_1^2}{a^2 b^2} = 1$

$\boxed{\dfrac{xx_1}{a^2} - \dfrac{yy_1}{b^2} = 1}$

If the point is $P(a\sec\theta, b\tan\theta)$ then $\dfrac{xa\sec\theta}{a^2} - \dfrac{yb\tan\theta}{b^2} = 1$
multiplying by $ab\cos\theta$ each term $x(\sec\theta)b\cos\theta - y(\tan\theta)a\cos\theta = ab\cos\theta$

$\boxed{bx - ay\sin\theta - ab\cos\theta = 0}$

Asymptotes to the Hyperbola
$\dfrac{x^2}{a^2} - \dfrac{y^2}{b^2} = 1$

The hyperbola is given by $b^2 x^2 - a^2 y^2 = a^2 b^2$ and the condition that $y = mx + c$ is tangent to the hyperbola may be found.

$b^2 x^2 - a^2(mx + c)^2 = a^2 b^2$

$b^2 x^2 - a^2 m^2 x^2 - a^2 c^2 - 2mxca^2 - a^2 b^2 = 0$

$x^2(b^2 - a^2 m^2) - 2mca^2 x - (a^2 c^2 + a^2 b^2) = 0 \quad \ldots (1)$

The straight line, $y = mx + c$, is tangent when the discriminant of equation (1) is zero.

$D = b^2 - 4ac$

$= (-2mca^2)^2 + 4(a^2 c^2 + a^2 b^2)(b^2 - a^2 m^2) = 0$

$4m^2 c^2 a^4 + 4a^2 c^2 b^2 + 4a^2 b^4 - 4a^4 c^2 m^2 - 4a^4 b^2 m^2 = 0$

$4a^2 b^2 (c^2 + b^2 - a^2 m^2) = 0$

$\boxed{c^2 = a^2 m^2 - b^2}$

Solving equation (1) by putting $D = 0$, we have

$x = \dfrac{2a^2 cm}{2(b^2 - a^2 m^2)}$

$= \dfrac{a^2 cm}{b^2 - a^2 m^2}$ but

$c = \pm\sqrt{a^2 m^2 - b^2}$

$x = \dfrac{a^2 m}{-c}$

$x = \pm\dfrac{a^2 m}{\sqrt{a^2 m^2 - b^2}}$

when $m \to \pm\dfrac{b}{a}$, $c = 0$, $x \to \pm 8$ and therefore

$\boxed{y = \pm\dfrac{b}{a}x}$ are the asymptotes.

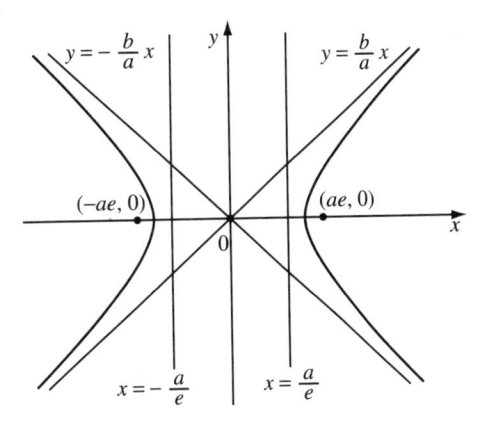

Fig. 6-I/64

WORKED EXAMPLE 27

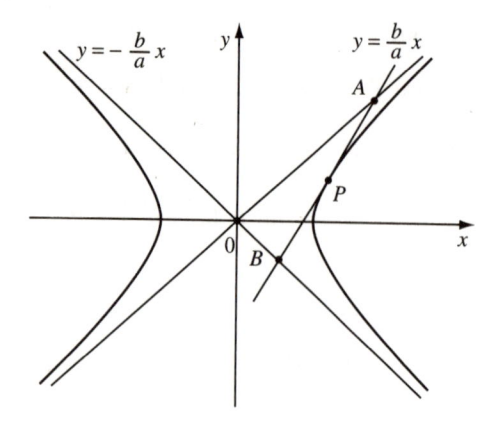

Fig. 6-I/65

The tangent at $P\left[\frac{a}{2}\left(p+\frac{1}{p}\right), \frac{b}{2}\left(p-\frac{1}{p}\right)\right]$ on the hyperbola $\frac{x^2}{a^2} - \frac{y^2}{b^2} = 1$ meets the asymptotes at A and B, find the locus of the mid-point of AB.

Solution 27

$x = \frac{a}{2}\left(p+\frac{1}{p}\right),$

$y = \frac{b}{2}\left(p-\frac{1}{p}\right)$

$\dfrac{\frac{a^2}{4}\left(p+\frac{1}{p}\right)^2}{a^2} - \dfrac{\frac{b^2}{4}\left(p-\frac{1}{p}\right)^2}{b^2}$

$= \frac{1}{4}\left(p^2 + 2 + \frac{1}{p^2}\right) - \frac{1}{4}\left(p^2 - 2 + \frac{1}{p^2}\right) = 1$

therefore $\left[\frac{a}{2}\left(p+\frac{1}{p}\right), \frac{b}{2}\left(p-\frac{1}{p}\right)\right]$ are the parametric coordinates of the hyperbola.

The tangent at P, is $\dfrac{xx_1}{a^2} - \dfrac{yy_1}{b^2} = 1,$

$\dfrac{x\frac{a}{2}\left(p+\frac{1}{p}\right)}{a^2} - \dfrac{y\frac{b}{2}\left(p-\frac{1}{p}\right)}{b^2} = 1$

$\boxed{\dfrac{x\left(p^2+1\right)}{a} - \dfrac{y\left(p^2-1\right)}{b} = 2p}$

This intersects $y = \frac{b}{a}x$ at A and

$y = -\frac{b}{a}x$ at B

$\dfrac{x\left(p^2+1\right)}{a} - \dfrac{b}{a}x\dfrac{\left(p^2-1\right)}{b} = 2p$

$xp^2 + x - xp^2 + x = 2ap$

$\boxed{x = ap}$

$y = \frac{b}{a}ap = bp$

$\boxed{y = bp}$

$A(ap, bp)$. The tangent intersects $ys = -\frac{b}{a}x$ at B

$\dfrac{x\left(p^2+1\right)}{a} + \dfrac{b}{a}x\dfrac{\left(p^2-1\right)}{b} = 2p$

$xp^2 + x + xp^2 - x = 2ap$

$2xp^2 = 2ap$

$\boxed{x = \frac{a}{p}}$

$y = -\frac{b}{a}x = -\frac{b}{a}\frac{a}{p} = -\frac{b}{p},$

$\boxed{y = -\frac{b}{p}}$

$B\left(\frac{a}{p}, -\frac{b}{p}\right)$. The coordinates of the mid point of AB are $\frac{1}{2}\left(ap + \frac{a}{p}\right)$ and $\frac{1}{2}\left(bp - \frac{b}{p}\right)$ which are the coordinates of the hyperbola, therefore the locus of the mid-point is the hyperbola $\frac{x^2}{a^2} - \frac{y^2}{b^2} = 1$.

WORKED EXAMPLE 28

For the following hyperbolas: Find
(a) the coordinates of the foci
(b) the coordinates of the vertices
(c) the equations of the asymptotes
(d) the eccentricities.

Sketch the curves.

(i) $\dfrac{x^2}{25} - \dfrac{y^2}{16} = 1$

(ii) $\dfrac{x^2}{3} - \dfrac{y^2}{2} = 1$

(iii) $\dfrac{y^2}{4} - \dfrac{x^2}{3} = 1$

(iv) $\dfrac{y^2}{2} - \dfrac{x^2}{1} = 1$

(v) $x^2 - y^2 = 1$

(vi) $y^2 - x^2 = 1$

(vii) $x^2 - y^2 = 4$

(viii) $\dfrac{x^2}{4^2} - \dfrac{y^2}{2^2} = 1$

(ix) $\dfrac{y^2}{\left(\frac{5}{3}\right)^2} - \dfrac{x^2}{\left(\frac{5}{4}\right)^2} = 1$.

Solution 28

(i) The standard equation is $\dfrac{x^2}{a^2} - \dfrac{y^2}{b^2} = 1$ from which $a^2 = 25, b^2 = 16$ hence $a = 5, b = 4$.

(a) $b^2 = a^2 e^2 - a^2$, $a^2 e^2 = b^2 + a^2 = 16 + 25 = 41$,

$ae = \pm\sqrt{41}$, $S'\left(-\sqrt{41}, 0\right)$, $S\left(\sqrt{41}, 0\right)$

(b) $A(5, 0), A'(-5, 0)$

(c) $y = \pm\dfrac{b}{a}x, y = \pm\dfrac{4}{5}x$,

$\boxed{y = \dfrac{4}{5}x}$, $\boxed{y = -\dfrac{4}{5}x}$

(d) $e = \pm\dfrac{\sqrt{41}}{5}$.

(ii) (a) $a = \sqrt{3}, b = \sqrt{2}$,

$(ae)^2 = b^2 + a^2 = 2 + 3 = 5$,

$ae = \pm\sqrt{5}$

$S'(-\sqrt{5}, 0), S(\sqrt{5}, 0)$

(b) $A(\sqrt{3}, 0), A'(-\sqrt{3}, 0)$

(c) $y = \pm\dfrac{b}{a}x = \pm\dfrac{\sqrt{2}}{\sqrt{3}}x$

(d) $e = \pm\dfrac{\sqrt{5}}{\sqrt{3}}$.

(iii) The hyperbola $\dfrac{y^2}{4} - \dfrac{x^2}{3} = 1$ has its transverse axis on the y-axis and $a = 2, b = \sqrt{3}$.

(a) $b^2 = a^2(e^2 - 1), 3 = 4(e^2 - 1), e^2 = 1.75$,

$e = \pm\sqrt{1.75}$ the foci are $\left(0, \dfrac{2\sqrt{7}}{2}\right)$, $\left(0, -\dfrac{2\sqrt{7}}{2}\right)$

(b) The vertices are $(0, 2); (0, -2)$

(c) $y = \dfrac{2x}{\sqrt{3}}, y = -\dfrac{2x}{\sqrt{3}}$

(d) $e = \pm\dfrac{\sqrt{7}}{2}$.

(iv) (a) $\left(0, \pm\sqrt{3}\right)$

(b) $\left(0, \pm\sqrt{2}\right)$

(c) $y = \pm\sqrt{2}x$

(d) $e = \pm\dfrac{\sqrt{3}}{\sqrt{2}}$

(v) (a) $\left(\pm\sqrt{2}, 0\right)$

(b) $(\pm 1, 0)$

(c) $y = \pm x$

(d) $e = \pm\sqrt{2}$.

(vi) The hyperbola $y^2 - x^2 = 1$ has its transverse axis on the y-axis and $a = 1, b = 1$.

(a) $b^2 = a^2(e^2 - 1) = e^2 - 1 = 1, e = \pm\sqrt{2}$, the coordinates of the foci are $(0, \sqrt{2})$ and $(0, -\sqrt{2})$.

(b) The vertices are $(0, 1)$ and $(0, -1)$

(c) $y = x, y = -x$

(d) $e = \pm\sqrt{2}$.

(vii) (a) $\left(\pm 2\sqrt{2}, 0\right)$

(b) $(\pm 2, 0)$

(c) $y = \pm x$

(d) $e = \pm\sqrt{2}$.

(viii) (a) $\left(\pm 2\sqrt{5}, 0\right)$

(b) $(\pm 4, 0)$

(c) $y = \pm\dfrac{x}{2}$

(d) $e = \pm\dfrac{\sqrt{5}}{2}$.

(ix) (a) $\left(0, \pm\dfrac{25}{12}\right)$

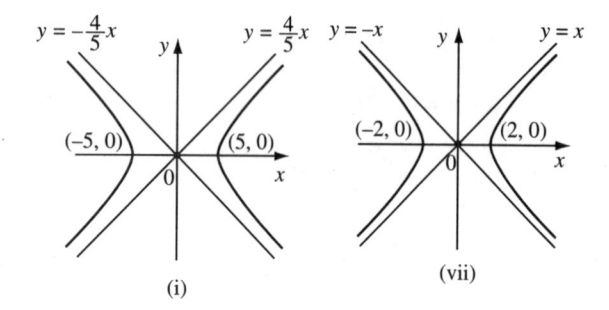

(b) $\left(0, \pm\dfrac{5}{3}\right)$

(c) $y = \pm\dfrac{4}{3}x$

(d) $e = \pm\dfrac{5}{4}$.

Fig. 6-I/66

Fig. 6-I/67

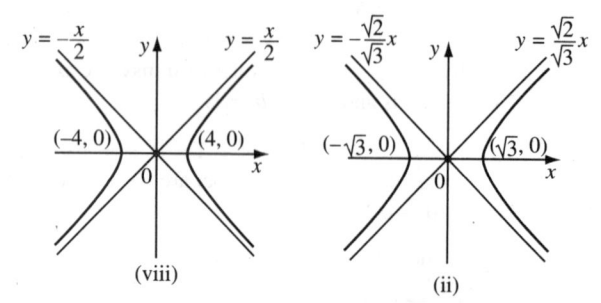

Fig. 6-I/68

Fig. 6-I/69

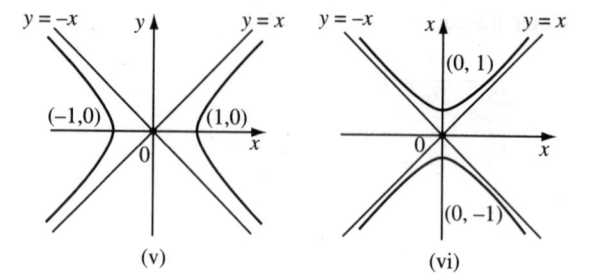

Fig. 6-I/70

Fig. 6-I/71

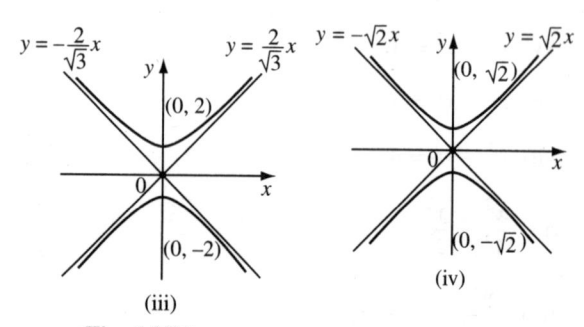

Fig. 6-I/72

Fig. 6-I/73

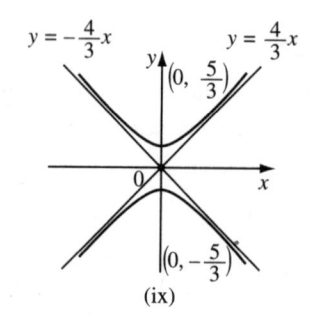

Fig. 6-I/74

The Rectangular Hyperbola

When the asymptotes are perpendicular to each other the hyperbola is called rectangular. The asymptotes are perpendicular when the gradients $m_1 m_2 = -1$

$$\left(-\dfrac{b}{a}\right)\left(\dfrac{b}{a}\right) = -1 = m_1 m_2 \qquad b^2 = a^2, b = a$$

Hence $\boxed{x^2 - y^2 = a^2}$

and the asymptotes are $y = x$ and $y = -x$ or $y - x = 0$ or $y + x = 0$ respectively.

The eccentricity is given by $a^2 = a^2(e^2 - 1)$, $1 = e^2 - 1$, $e^2 = 2$, $\boxed{e = \pm\sqrt{2}}$

WORKED EXAMPLE 29

Determine the eccentricities, the equations of the asymptote and the foci of the following hyperbolas:

(i) $b^2x^2 - a^2y^2 = a^2b^2$

(ii) $x^2 - 3y^2 = 9$

(iii) $x^2 - y^2 = 16$.

Solution 29

(i) $b^2 = a^2(e^2 - 1)$ hence $\dfrac{b^2}{a^2 + 1} = e^2$,

$$e = \pm\dfrac{\sqrt{b^2 + a^2}}{a}$$

$y = \dfrac{b}{a}x$ and $y = -\dfrac{b}{a}x$ are the asymptotes, the foci $\left(\pm\sqrt{a^2 + b^2}, 0\right)$.

(ii) $\dfrac{x^2}{3^2} - \dfrac{y^2}{\left(\sqrt{3}\right)^2} = 1$

$$a = 3, b = \sqrt{3}$$

$$b^2 = a^2(e^2 - 1)$$

$$\dfrac{b^2}{a^2} = e^2 - 1,$$

$$\dfrac{b^2 + a^2}{a^2} = e^2,$$

$$e = \pm\sqrt{\dfrac{3+9}{9}}$$

$$= \pm\sqrt{\dfrac{4}{3}}$$

$$= \pm\dfrac{2}{\sqrt{3}}.$$

The asymptotes are $y = \pm\dfrac{\sqrt{3}}{3}x$ and the foci $\left(\pm 3\left(\dfrac{2}{\sqrt{3}}\right), 0\right) = \left(\pm 2\sqrt{3}, 0\right)$.

(iii) $a^2 = a^2(e^2 - 1)$, $e^2 = 2$, $e = \pm\sqrt{2}$ the asymptotes $y = \pm x$ and the foci $\left(\pm 4\sqrt{2}, 0\right)$.

WORKED EXAMPLE 30

Determine the equation of the chord joining the points $P\left(4\sqrt{2}, 2\right)$ and $Q(5, 1.5)$ on the hyperbola $x^2 - 4y^2 = 16$. Determine also the coordinates of the line intersecting the axes.

Solution 30

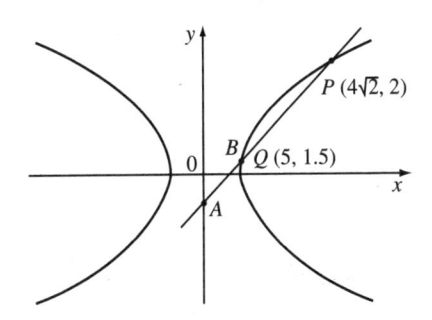

Fig. 6-I/75

$$\dfrac{2 - 1.5}{4\sqrt{2} - 5} = \dfrac{y - 1.5}{x - 5} \text{ or}$$

$$0.5(x - 5) = \left(4\sqrt{2} - 5\right)(y - 1.5).$$

If $x = 0$

$$-2.5 = \left(4\sqrt{2} - 5\right)(y - 1.5)$$

$$= 4\sqrt{2}y - 1.5 \times 4\sqrt{2} - 5y + 7.5$$

$$y\left(5 - 4\sqrt{2}\right) = 10 - 6\sqrt{2} \Rightarrow y = \dfrac{10 - 6\sqrt{2}}{5 - 4\sqrt{2}}$$

$$y = -2.31 \quad \boxed{A = (0, -2.31)}$$

If $y = 0$,

$$0.5x - 2.5 = \left(4\sqrt{2} - 5\right)(-1.5)$$

$$= -6\sqrt{2} + 7.5$$

$$0.5x = -6\sqrt{2} + 10$$

$$x = -12\sqrt{2} + 20.$$

$$x = 3.03 \quad \boxed{B = (3.03, 0)}.$$

Rectangular Hyperbola

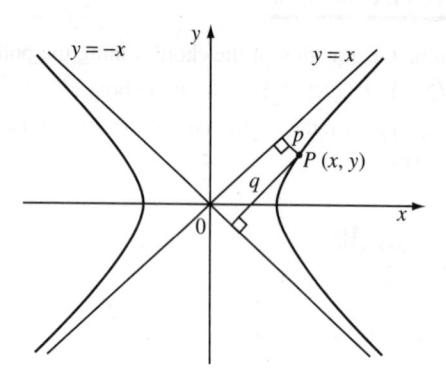

Fig. 6-I/76

Let p and q be the perpendicular distances of point $P(x, y)$ from $y = x$ and $y = -x$ as shown in Fig. 6-I/76.

Employing the formula $p = \pm \frac{ax_1+by_1+c}{\sqrt{a^2+b^2}}$ where p is the perpendicular distance from a point (x_1, y_1) to the line $ax + by + c = 0$

$$p = \pm \frac{x-y}{\sqrt{2}} \text{ and } q = \pm \frac{x+y}{\sqrt{2}}$$

$$pq = \pm \frac{x^2 - y^2}{2} \text{ and since } x^2 - y^2 = a^2,$$

$$pq = \pm \frac{a^2}{2}$$

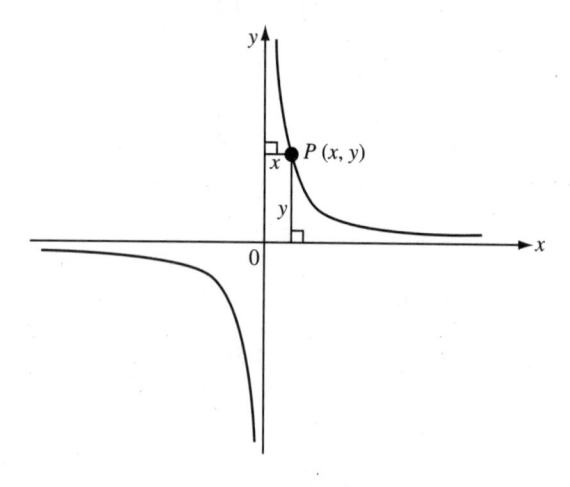

Fig. 6-I/77

Fig. 6-I/77 shows the rectangular hyperbola referred to the x and y axes as their asymptotes $\boxed{xy = c^2}$ where $c > 0$.

Conjugate Axis

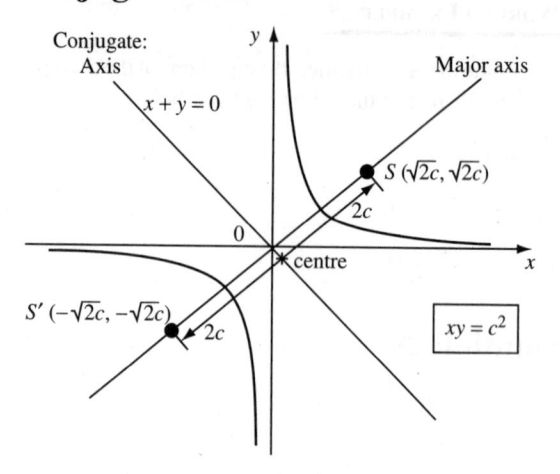

Fig. 6-I/78

$a^2 = a^2(e^2 - 1)$, $e^2 = 2$, $e = \pm\sqrt{2}$ and the foci are $\left(\pm\sqrt{2}a, 0\right)$ for the rectangular hyperbola $x^2 - y^2 = a^2$. Let $\frac{a^2}{2} = c^2$, hence $a = \sqrt{2}c$. Therefore the foci are $\left(\pm\sqrt{2}\sqrt{2}c, 0\right)$ from centre of the major axis. Therefore the coordinates of the foci of $xy = c^2$ are $\left(\sqrt{2}c, \sqrt{2}c\right)$ and $\left(-\sqrt{2}c, -\sqrt{2}c\right)$.

The line $x + y = 0$ do not cut the $xy = c^2$; it is called the conjugate axis.

The Equation of the Tangent to the Hyperbola $xy = c^2$ at the Point $P(x_1, y_1)$

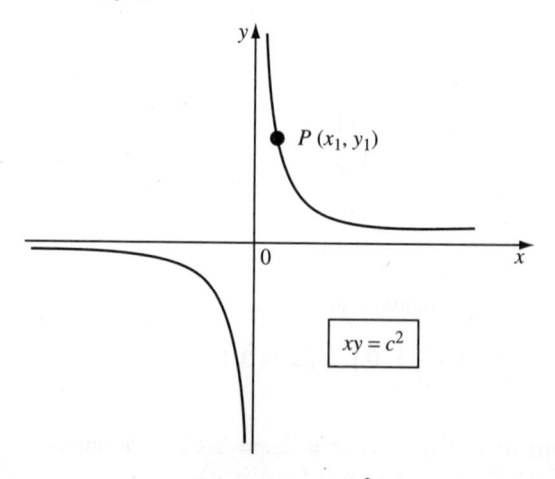

Fig. 6-I/79 $xy = c^2$

$xy = c^2$ ... (1)

Differentiating (1) with respect to x

$1 \cdot y + x\dfrac{dy}{dx} = 0,$

$\dfrac{dy}{dx} = -\dfrac{y}{x}$, the gradient at any point. At $P(x_1, y_1)$

$\dfrac{dy}{dx} = -\dfrac{y_1}{x_1}$, the equation of the tangent is $y = -\dfrac{y_1}{x_1}x + k$ where k is a constant, the y intercept of a straight line. This passes through $P(x_1, y_1)$.

$k = y_1 + \dfrac{y_1}{x_1}x_1 = 2y_1$

$y = -\dfrac{.y_1}{x_1}x + 2y_1$ or $yx_1 + xy_1 = 2x_1y_1 = 2c^2$

$\boxed{yx_1 + xy_1 = 2c^2}$

The Parametric Equations of the Rectangular Hyperbola $xy = c^2$

Equation of the Tangent at $P\left(ct, \dfrac{c}{t}\right)$

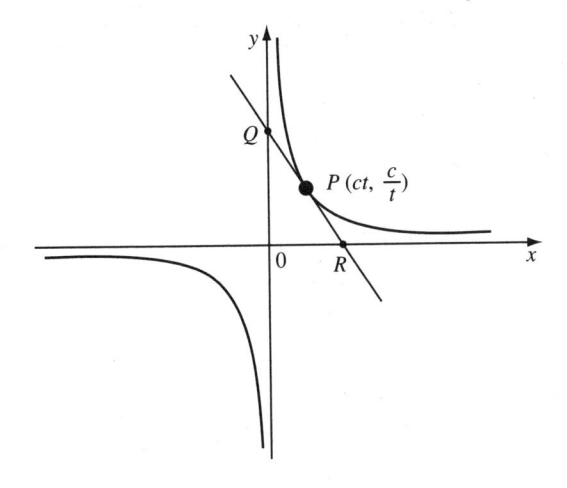

Fig. 6-I/80

$x = ct, \dfrac{dx}{dt} = c;$

$y = \dfrac{c}{t} = ct^{-1},$

$\dfrac{dy}{dt} = -ct^{-2} = -\dfrac{c}{t^2}$

$\dfrac{dy}{dx} = \dfrac{\frac{dy}{dt}}{\frac{dx}{dt}} = \dfrac{-\frac{c}{t^2}}{c} = -\dfrac{1}{t^2},$

the gradient of the tangent at $P\left(ct, \dfrac{c}{t}\right)$. The equation of the tangent is

$y = -\dfrac{1}{t^2}x + k.$

This passes through $P\left(ct, \dfrac{c}{t}\right)$,

$k = \dfrac{c}{t} + \dfrac{1}{t^2}ct = \dfrac{c}{t} + \dfrac{c}{t} = \dfrac{2c}{t}$

$y = -\dfrac{1}{t^2}x + \dfrac{2c}{t}.$

The equation of the tangent $\boxed{x + t^2y - 2ct = 0}$ ... (1)

WORKED EXAMPLE 31

Referring to Fig. 6-I/80, determine the coordinates of Q and R.

Solution 31

From equation (1) we have:

If $x = 0$, $y = \dfrac{2c}{t}$; if $y = 0, x = 2ct.$

Therefore, $Q\left(0, \dfrac{2c}{t}\right)$ and $R(2ct, 0)$ in Fig. 6-I/80.

WORKED EXAMPLE 32

Determine the locus of the mid-point of the tangent QR.

Solution 32

The coordinates of the mid-point of QR are given as $M\left(\dfrac{0 + 2ct}{2}, \left(\dfrac{2c}{t} + 0\right)\dfrac{1}{2}\right) x = ct, y = \dfrac{c}{t}.$

Eliminating the parameter from $x = ct, y = \dfrac{c}{t}$, we have

$xy = ct \cdot \dfrac{c}{t} = c^2$

$\boxed{xy = c^2}$

The locus is the hyperbola itself.

The Equation of the Normal to the Hyperbola $xy = c^2$ to the Point $P\left(ct, \frac{c}{t}\right)$

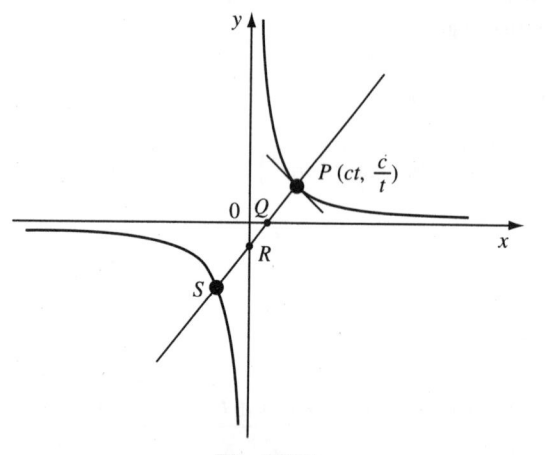

Fig. 6-I/81

$xy = c^2$, differentiating with respect to x $1 \cdot y + x\frac{dy}{dx} = 0$, $\frac{dy}{dx} = -\frac{y}{x}$, the gradient of the tangent at any point, the gradient of the normal at any point is $\frac{dy}{dx} = \frac{x}{y}$. At $P\left(ct, \frac{c}{t}\right)$, $\frac{dy}{dx} = \frac{ct}{\frac{c}{t}} = t^2$. The equation of the normal is $y = t^2 x + k$, this passes through $P\left(ct, \frac{c}{t}\right)$, $\frac{c}{t} = t^2 ct + k$, $k = \frac{c}{t} - ct^3$ $y = t^2 x + \frac{c}{t} - ct^3$, therefore the equation of the normal is

$$\boxed{t^2 x - y - ct^3 + \frac{c}{t} = 0} \qquad \ldots (2)$$

WORKED EXAMPLE 33

The normal at $P\left(ct, \frac{c}{t}\right)$ of the hyperbola $xy = c^2$ intersects the axes at Q and R, referring to Fig. 6-I/81, find these coordinates of Q, R and S where this normal meets the curve again.

Solution 33

If $x = 0$ in equation (2), $y = -ct^3 + \frac{c}{t}$.

If $y = 0$ in equation (2), $x = ct - \frac{c}{t^3}$.

The coordinates of Q and R are $Q\left(ct - \frac{c}{t^3}, 0\right)$ and $R\left(0, -ct^3 + \frac{c}{t}\right)$.

Solving $xy = c^2$ and $t^2 x - y - ct^3 + \frac{c}{t} = 0$ simultaneously, we have

$$t^2 x - \frac{c^2}{x} - ct^3 + \frac{c}{t} = 0$$

$$t^2 x^2 - xct^3 + x\frac{c}{t} - c^2 = 0.$$

This is a quadratic equation in x, solving it we have

$$x = \frac{\left(ct^3 - \frac{c}{t}\right) \pm \sqrt{\left(ct^3 - \frac{c}{t}\right)^2 + 4c^2 t^2}}{2t^2}$$

$$x = \frac{ct}{2} - \frac{c}{2t^3} \pm \left(ct^3 + \frac{c}{t}\right)\frac{1}{2t^2}$$

$$x = \frac{ct}{2} - \frac{c}{2t^3} + \frac{ct}{2} + \frac{c}{2t^3} = ct \qquad \ldots (1)$$

or $x = \frac{ct}{2} - \frac{c}{2t^3} - \frac{ct}{2} - \frac{c}{2t^3} = -\frac{c}{t^3} \ldots \qquad \ldots (2)$

Substituting for y, we have

$$y = \frac{c^2}{x} = \frac{c^2}{ct} = \frac{c}{t}$$

or $y = \frac{c^2}{\frac{-c^3}{t}} = -ct^3$.

The coordinates of the meet of the normal with the curve are $S\left(-\frac{c}{t^3}, -ct^3\right)$.

WORKED EXAMPLE 34

Find the locus of the point of intersection R, of the two normals at $P\left(cp, \frac{c}{p}\right)$ and $Q\left(cq, \frac{c}{q}\right)$ of the rectangular hyperbola $xy = c^2$, given the condition that $pq = -1$.

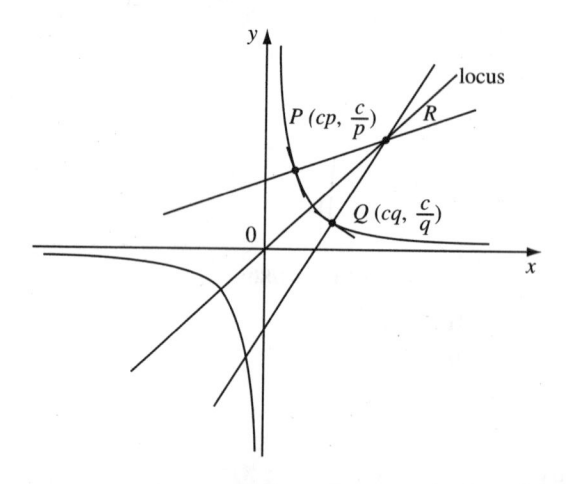

Fig. 6-I/82

Solution 34

The equation of the normal at $\left(ct, \dfrac{c}{t}\right)$ is $t^2 x - y - ct^3 + \dfrac{c}{t} = 0$.

The equations of the normals at P and Q are $p^2 x - y - cp^3 + \dfrac{c}{p} = 0$, and $q^2 x - y - cq^3 + \dfrac{c}{q} = 0$. Solving these equations simultaneously we have $p^2 x - q^2 x - cp^3 + cq^3 + \dfrac{c}{p} - \dfrac{c}{q} = 0$

$$x(p^2 - q^2) = c(p^3 - q^3) + c\left(\dfrac{1}{q} - \dfrac{1}{p}\right)$$

$$x = c\dfrac{p^3 - q^3}{p^2 - q^2} + c\dfrac{p - q}{pq(p^2 - q^2)}$$

$$x = c\dfrac{(p^2 + q^2 + pq)}{p + q} + \dfrac{c}{pq(p + q)}$$

$$x = c\dfrac{[(p+q)^2 - 2pq + pq]}{p + q} + \dfrac{c}{pq(p + q)}$$

Substituting $pq = -1$

$$x = c(p + q) + \dfrac{c}{p + q} - \dfrac{c}{p + q} = c(p + q),$$

$$\boxed{x = c(p + q)}$$

Substituting x in

$$p^2 x - y - cp^3 + \dfrac{c}{p} = 0$$

$$y = p^2 x - cp^3 + \dfrac{c}{p}$$

$$= p^2 c(p + q) - cp^3 + \dfrac{c}{p}$$

$$= cp^2 q + \dfrac{c}{p}$$

$$y = c\left(\dfrac{p^3 q + 1}{p}\right)$$

$$= c\left(\dfrac{pqp^2 + 1}{p}\right)$$

$$= c\dfrac{(1 - p^2)}{p}$$

$$R\left[c(p + q), c\dfrac{(1 - p^2)}{p}\right].$$

The locus is found by eliminating p and q, in $x = c(p + q)$ and $y = c\dfrac{(1 - p^2)}{p}$ and using the condition $pq = -1$.

$$x = c(p + q) = c\left(p - \dfrac{1}{p}\right) = \dfrac{c(p^2 - 1)}{p}$$

$$y = \dfrac{c(1 - p^2)}{p}, \text{ therefore } \boxed{y = x} \text{ is the locus.}$$

The Equation of the Chord Joining the Points $P\left(cp, \dfrac{c}{p}\right)$ and $Q\left(cq, \dfrac{c}{q}\right)$

Equation of a straight line passing through PQ, the chord.

$$\dfrac{\dfrac{c}{p} - \dfrac{c}{q}}{cp - cq} = \dfrac{y - \dfrac{c}{p}}{x - cq}$$

$$\dfrac{\dfrac{(q-p)}{pq}}{(p - q)} = \dfrac{y - \dfrac{c}{p}}{x - cq},$$

$$\dfrac{-(x - cq)}{pq} = y - \dfrac{c}{p}$$

$$-x + cq = pqy - cq \text{ or } \boxed{pqy + x - 2cq = 0}$$

If Q approaches P, then $q \to p$ $\boxed{p^2 y + x - 2cp = 0}$ is the equation of the tangent at P if P approaches Q, the $p \to q$ $\boxed{q^2 y + x - 2cq = 0}$ is the equation of the tangent at Q.

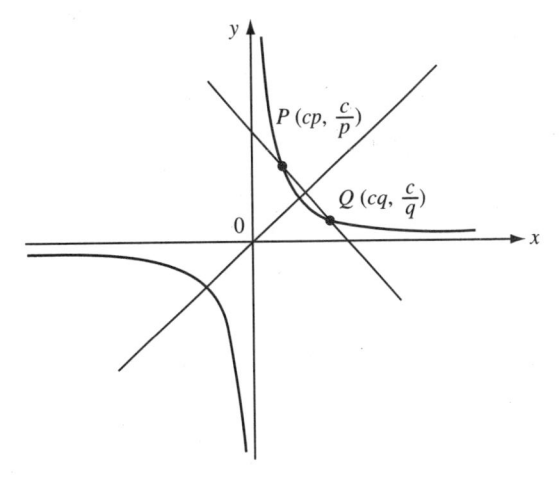

Fig. 6-I/83

The Locus of the Point $x = ct$, $y = \dfrac{c}{t}$

The rectangular hyperbola is the locus of a point $P(x, y)$ determined by the parametric equations

$$\boxed{x = ct} \quad \ldots (1) \qquad \boxed{y = \dfrac{c}{t}} \quad \ldots (2)$$

where c is a positive constant. Multiplying equation (1) and (2) $\boxed{xy = c^2}$

WORKED EXAMPLE 35

Draw the graphs $xy = 1$, if (i) $x > 0$, (ii) $x < 0$.

Solution 35

(i)

x	0.001	0.01	0.1	1	10	100	1000
y	1000	100	10	1	0.1	0.01	0.001

(ii)

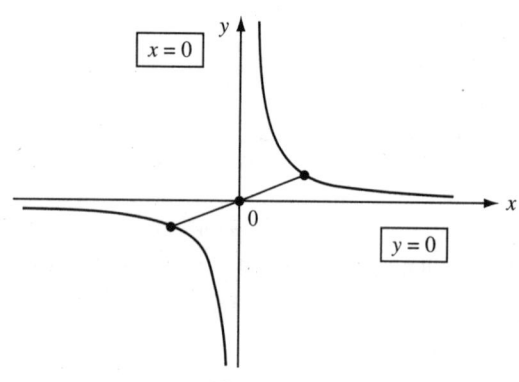

x	-0.001	-0.01	-0.1	-1	-10	-100	-1000
y	-1000	-100	-10	-1	-0.1	-0.01	-0.001

Fig. 6-I/84

The Properties of the Rectangular Hyperbola $xy = c^2$

O is the centre of the curve and any chord through O is called a diameter. The curve $xy = c^2$ is symmetrical about its centre, which is also the point of intersection of the asymptotes (the x-axis and y-axis, $y = 0$ and $x = 0$ respectively).

The asymptotes, $y \to 0$, $x \to +\infty$; $x \to 0+$, $y \to +\infty$, $y = 0-$, when $x \to -\infty$ and $x \to 0-$ when $y \to -\infty$, where the symbols $0+$ and $0-$ mean approach to zero through positive and negative values respectively.

$x + y = 0$ and $x - y = 0$ are the axes of symmetry or the axes of the rectangular hyperbola $xy = c^2$.

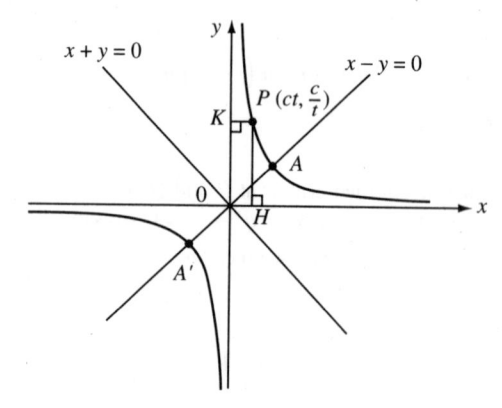

Fig. 6-I/85

A and A' are called the vertices. AA' is the transverse axis of the rectangular hyperbola.

$AA' = 2a$, $a^2 = OA^2 = 2c^2$.

The hyperbola is called rectangular because the asymptotes are at right angles.

$PK \cdot PH = ct \cdot \dfrac{c}{t} = c^2 =$ area of the rectangle $OKPH$.

Area between the Tangent and the Axes

WORKED EXAMPLE

The tangent at $P\left(ct, \dfrac{c}{t}\right)$ cuts x and y axes at the points A and B, find their coordinates. Determine the area of the triangle APB.

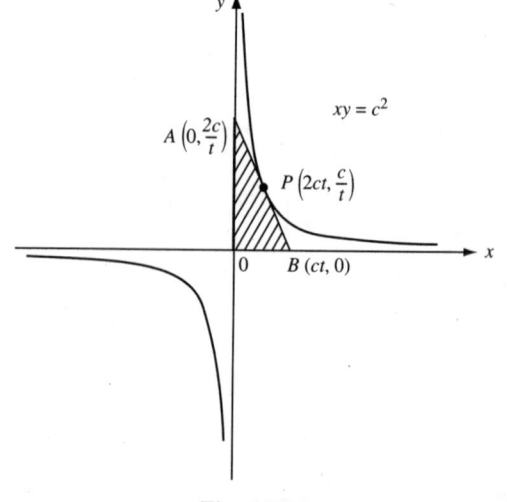

Fig. 6-I/86

Solution

The tangent at P is given as $yt^2 + x = 2ct$, when $x = 0$, $y = \frac{2c}{t}$ and when $y = 0$, $x = 2ct$. The coordinates of A and B are $A\left(0, \frac{2c}{t}\right)$ and $B(2ct, 0)$.

The area of the triangle APB is given as $\frac{1}{2} \frac{2c}{t} \cdot 2ct = 2c^2$.

The Equation of a Chord through $P_1\left(ct_1, \frac{c}{t_1}\right)$ and $P_2\left(ct_2, \frac{c}{t_2}\right)$

$$m = \frac{\frac{c}{t_1} - \frac{c}{t_2}}{ct_1 - ct_2} = \frac{c\frac{(t_2 - t_1)}{t_1 t_2}}{c(t_1 - t_2)}$$

$$= -\frac{1}{t_1 t_2} \text{ where } t_1 \neq t_2$$

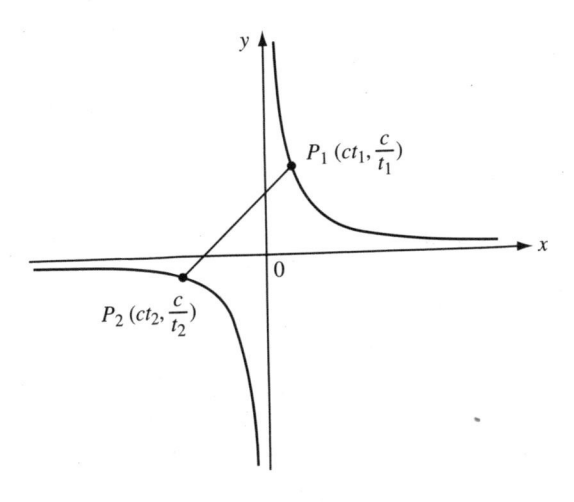

Fig. 6-I/87

$y = mx + b = -\frac{1}{t_1 t_2} x + b$, this passes through P_1 or P_2.

Through P_1, $\frac{c}{t_1} = -\frac{1}{t_1 t_2} \cdot ct_1 + b$,

$$b = \frac{c}{t_1} + \frac{c}{t_2}$$

$$y = -\frac{1}{t_1 t_2} x + \frac{c}{t_1} + \frac{c}{t_2}$$

$$yt_1 t_2 + x = ct_2 + ct_1$$

$$\boxed{x + t_1 t_2 y = c(t_1 + t_2)}$$

The Equation of a Rectangular Hyperbola Referred to Axes Parallel to the Asymptotes through the Point (h, k) is

$$\boxed{(x - h)(y - k) = c^2}$$

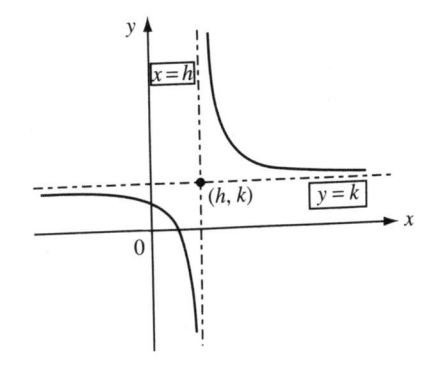

Fig. 6-I/87

The point $\left(cp + h, \frac{c}{p} + k\right)$ lies on the curve $(x - h)(y - k) = c^2$ since substituting $x = cp + h$, and $y = \frac{c}{p} + k$ in the equation verifies it, $(cp + h - h)\left(\frac{c}{p} + k - k\right) = c^2$.

WORKED EXAMPLE

Show that the point $\left(2p + 1, \frac{2}{p} - 2\right)$ lies on the curve $(x - 1)(y + 2) = 2^2$ for all values of p.

Solution

Substituting $x = 2p + 1$, $y = \frac{2}{p} - 2$ in the equation $(x-1)(y+2) = 2^2$, we have $(2p+1-1)(2p-2+2) = 2p\left(\frac{2}{p}\right) = 2^2$.

WORKED EXAMPLE

Show that the point $\left(2p - 5, \frac{2}{p} + 3\right)$ lies on the curve $(x + 5)(y - 3) = 4$.

Solution

Substituting $x = 2p - 5$, $y = \frac{2}{p} + 3$ in $(x+5)(y-3) = 4$ verifies it, $(2p - 5 + 5)\left(\frac{2}{p} + 3 - 3\right) = 4$.

WORKED EXAMPLE 36

Show that the equation of the normal to the hyperbola $xy = c^2$ at the point

$P_1(x_1, y_1)$ is $\boxed{xx_1 - yy_1 = x_1^2 - y_1^2}$

Solution 36

The gradient of the normal at $P_1(x_1, y_1)$, $\dfrac{dy}{dx} = \dfrac{x_1}{y_1}$

$y = \dfrac{x_1}{y_1}x + b$, at $P_1(x_1, y_1)$, $y_1 = \dfrac{x_1}{y_1}x_1 + b$

$b = \dfrac{y_1^2 - x_1^2}{y_1} \quad y = \dfrac{x_1}{y_1}x + \dfrac{y_1^2 - x_1^2}{y_1}$

$yy_1 = xx_1 + y_1^2 - x_1^2$

$\boxed{xx_1 - yy_1 = x_1^2 - y_1^2}$

The Diameter of the Rectangular Hyperbola $xy = c^2$

Any chord that passes through the centre, O, is called diameter.

WORKED EXAMPLE 37

The normal at $P\left(cp, \dfrac{c}{p}\right)$ meets the curve again at $Q\left(cq, \dfrac{c}{q}\right)$. Determine the coordinates of Q in terms of p. QR is the diameter through Q of the hyperbola. Determine the locus of the mid-point of PR as p varies.

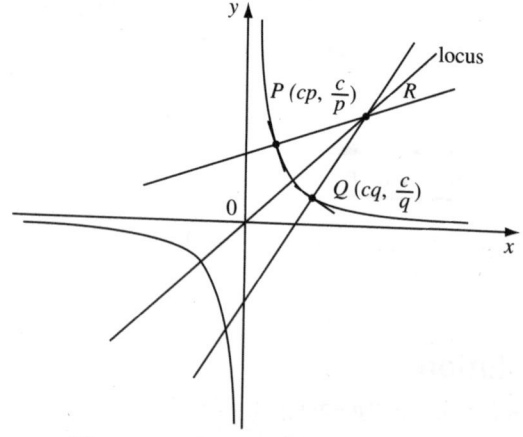

Fig. 6-I/88 The diameter of a hyperbola

Solution 37

The equation of the normal at $T\left(ct, \dfrac{c}{t}\right)$ is given $t^2 x - y - ct^3 + \dfrac{c}{t} = 0$.

The equation of the normal at $P\left(cp, \dfrac{c}{p}\right)$ is

$p^2 x - y - cp^3 + \dfrac{c}{p} = 0$ ...(1)

and $xy = c^2$ ...(2)

Solving the simultaneous equation (1) and (2)

we have $P\left(cp, \dfrac{c}{p}\right)$ and $Q\left(-\dfrac{c}{p^3}, -cp^3\right)$

where $q = -\dfrac{1}{p^3}$.

Equation QR or OQ, $y = \dfrac{-cp^3}{-\dfrac{c}{p^3}}x = p^6 x$

$\boxed{y = p^6 x}$ equation of the diameter. To find the coordinate of R, solve $y = p^6 x$ and $xy = c^2$ simultaneously.

$y = p^6 x = \dfrac{c^2}{x} \Rightarrow x^2 = \dfrac{c^2}{p^6}$,

$x = \pm\dfrac{c}{p^3}$ and $y = p^6\left(\pm\dfrac{c}{p^3}\right) = \pm cp^3$,

therefore if $Q\left(-\dfrac{c}{p^3}, -cp^3\right)$ then $R\left(\dfrac{c}{p^3}, cp^3\right)$. The mid point M has coordinates $M\left[\left(cp + \dfrac{c}{p^3}\right)\dfrac{1}{2}, \left(\dfrac{c}{p} + cp^3\right)\dfrac{1}{2}\right]$.

$x = \dfrac{1}{2}\left(cp + \dfrac{c}{p^3}\right)$ and $y = \dfrac{1}{2}\left(\dfrac{c}{p} + cp^3\right)$

$2x = cp + \dfrac{c}{p^3}$, $2y = \dfrac{c}{p} + cp^3$

$2x = \dfrac{c(p^4 + 1)}{p^3}$, $2y = \dfrac{c(1 + p^4)}{p}$

$2xp^3 = 2yp$, $p^2 = \dfrac{y}{x}$

$2xp^3 = c(p^4 + 1)$

$\pm 2x\left(\dfrac{y}{x}\right)\sqrt{\dfrac{y}{x}} = c\left(\dfrac{y^2}{x^2} + 1\right)$.

Squaring up both sides

$4x^2\dfrac{y^2}{x^2}\dfrac{y}{x} = \dfrac{c^2(x^2 + y^2)^2}{x^4}$

$\boxed{4x^3 y^3 = c^2(x^2 + y^2)^2}$

Conjugate Hyperbolas

The rectangular hyperbola $x^2 - y^2 = a^2$ with asymptotes $x + y = 0$ and $x - y = 0$ and the rectangular hyperbola $x^2 - y^2 = -a^2$ with asymptotes $x + y = 0$ and $x - y = 0$ are given in Fig. 6-I/89.

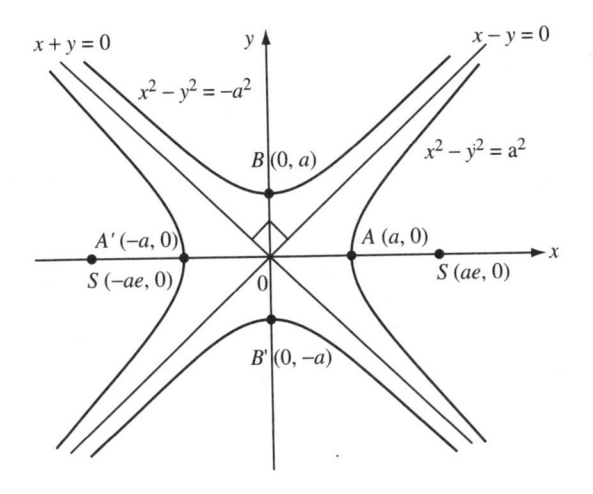

Fig. 6-I/89

$x^2 - y^2 = a^2$ and $x^2 - y^2 = -a^2$ are called <u>Conjugate hyperbolas</u>, where the transverse axis $A'OA$ and the conjugate axis $B'OB$ of the rectangular hyperbola are taken as x-axis and y-axis. The vertices are $A(a, 0)$, $A'(-a, 0)$ for $x^2 - y^2 = a^2$, and $B(0, a)$ and $B'(0, -a)$ for $x^2 - y^2 = -a^2$.

The equations of conjugate rectangular hyperbolas are of the form $xy = c^2$, $xy = -c^2$.

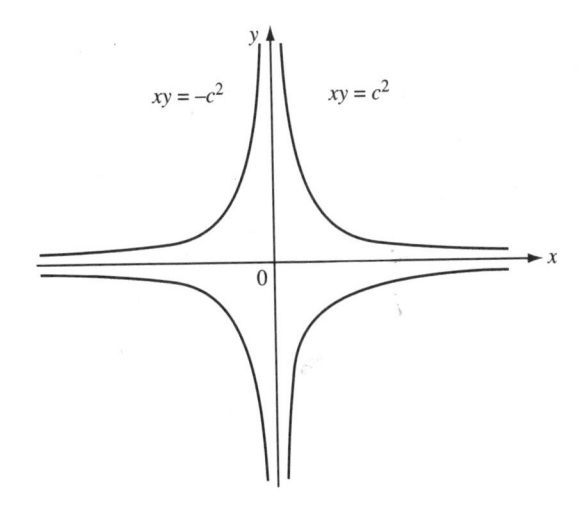

Fig. 6-I/90

WORKED EXAMPLE 38

Write down the coordinates of the centre and the equations of the asymptotes of the following rectangular hyperbolas:

(i) $xy = 9$

(ii) $xy = 5^2$

(iii) $(x - 1)(y - 2) = 4$

(iv) $(x + 2)(y - 5) = 25$

(v) $(x - h)(y - k) = k^2$.

Hence sketch the hyperbola.

Solution 38

(i) $xy = 9$, $C(0, 0)$, the coordinates of the centre and $x = 0$ and $y = 0$ the asymptotes

(ii) $xy = 5^2$, $C(0, 0)$, the coordinates of the centre and $x = 0$ and $y = 0$ the asymptotes

(iii) $(x - 1)(y - 2) = 4$, $C(1, 2)$, the coordinates of the centre and $x = 1$ and $y = 2$ the asymptotes

(iv) $(x + 2)(y - 5) = 25$, $C(-2, 5)$, the coordinates of the centre and $x = -2$ and $y = 5$ the asymptotes

(v) $(x - h)(y - k) = k^2$, $C(h, k)$, the coordinates of the centre and $x = h$ and $y = k$ the asymptotes.

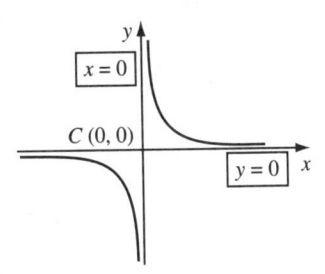

Fig. 6-I/91 (i)

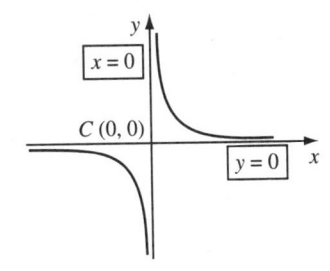

Fig. 6-I/92 (ii)

54 — GCE A level

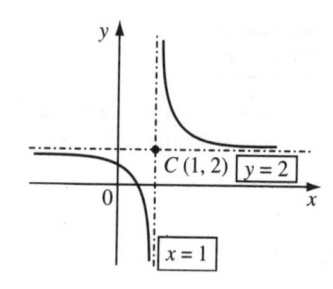

Fig. 6-I/93 (iii)

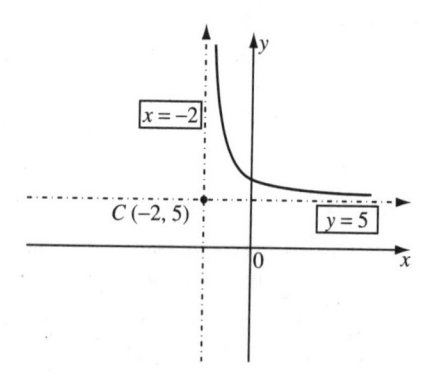

Fig. 6-I/94 (iv)

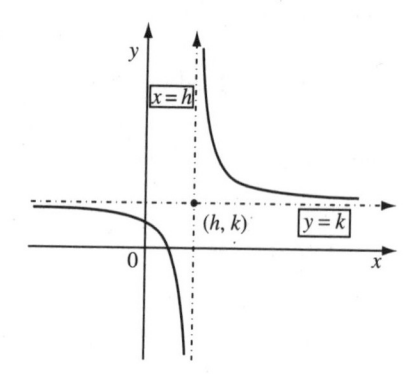

Fig. 6-I/95 (v)

WORKED EXAMPLE 39

Sketch the hyperbolas $xy = 1$ and $(x+1)(y-2) = 1$ and determine the coordinates of the points of intersection.

Solution 39

$xy = 1$ ... (1)

and $(x+1)(y-2) = 1$ ... (2)

Solving the equation (1) and (2) we have $xy = 1$, $xy + y - 2x - 2 = 1$

$$y - 2x - 2 = 0,$$

$$y = 2x + 2$$

$$x(2x + 2) = 1,$$

$$2x^2 + 2x - 1 = 0,$$

$$x = \frac{-2 \pm \sqrt{4 + 8}}{4}$$

$$x = -\frac{1}{2} \pm \frac{\sqrt{3}}{2},$$

$$x = -\frac{1}{2} + \frac{\sqrt{3}}{2}$$

or $x = -\dfrac{1}{2} - \dfrac{\sqrt{3}}{2}$

and the corresponding values of y are

$$y = 2\left(-\frac{1}{2} + \frac{\sqrt{3}}{2}\right) + 2 = \sqrt{3} + 1$$

or $y = 2\left(-\dfrac{1}{2} - \dfrac{\sqrt{3}}{2}\right) + 2 = -\sqrt{3} + 1.$

The coordinates at P and Q are

$$P\left(\frac{\sqrt{3} - 1}{2}, \sqrt{3} + 1\right) \quad Q\left(-\frac{\sqrt{3} + 1}{2}, -\sqrt{3} + 1\right).$$

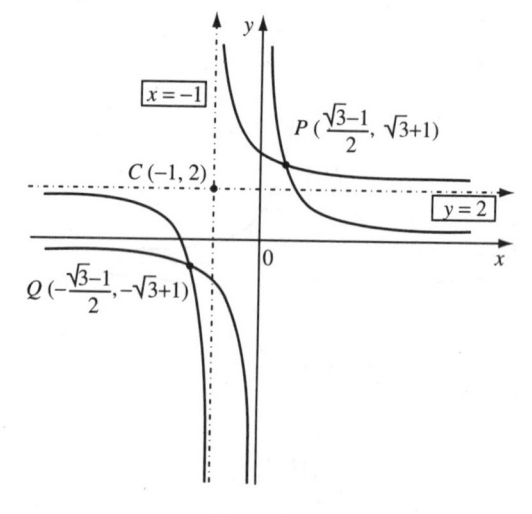

Fig. 6-I/96

The Hyperbola — 55

WORKED EXAMPLE 40

Find the equation of the normal to the rectangular hyperbola $xy = c^2$ at the point $P\left(cp, \frac{c}{p}\right)$. Show that the area of the triangle formed by this normal and the axes of coordinates is $c^2\left(1 - \frac{1}{2}p^4 - \frac{1}{2p^4}\right)$. Find the coordinates of the centroid of this triangle and hence its locus.

Solution 40

The equation of the normal is $p^2 x - y - cp^3 + \dfrac{c}{p} = 0$.

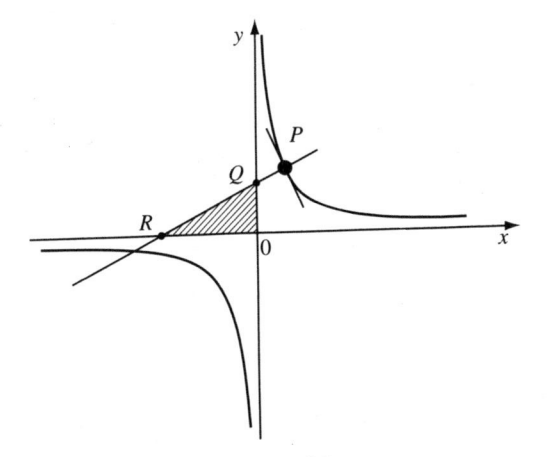

Fig. 6-I/97

The normal intersects the y-axis at Q when $x = 0$, $y = \frac{c}{p} - cp^3$, $Q\left(0, \frac{c}{p} - cp^3\right)$ and intersects the x-axis at R when $y = 0$, $x = cp - \frac{c}{p^3}$, $R\left(cp - \frac{c}{p^3}, 0\right)$.

The area of the triangle OQR is

$$\frac{1}{2}\left(\frac{c}{p} - cp^3\right)\left(cp - \frac{c}{p^3}\right)$$

$$= \frac{1}{2}c^2 - \frac{1}{2}\frac{c^2}{p^4} - \frac{1}{2}c^2 p^4 + \frac{1}{2}c^2$$

$$= c^2 - \frac{1}{2}c^2\left(p^4 + \frac{1}{p^4}\right)$$

$$\boxed{\text{Area} = c^2\left(1 - \frac{1}{2}p^4 - \frac{1}{2p^4}\right)}$$

The coordinates of the centroid are

$$\left[\frac{cp}{3}\left(1 - \frac{1}{p^4}\right), \frac{c}{3p}\left(1 - p^4\right)\right].$$

The locus is found by eliminating the parameter between the equations

$$x = \frac{cp}{3}\left(1 - \frac{1}{p^4}\right) \text{ and } y = \frac{c}{3p}\left(1 - p^4\right)$$

$$x = \frac{c}{3}\left[p - \frac{1}{p^3}\right] \text{ and } y = \frac{c}{3}\left[\frac{1}{p} - p^3\right]$$

$$\frac{x}{y} = \frac{(p^4 - 1) \times p}{(1 - p^4) \times p^3} = -\frac{1}{p^2}$$

$$\frac{x^2}{y^2} = \frac{1}{p^4} \text{ or }$$

$$p^4 = \frac{y^2}{x^2}$$

$$y^2 = \frac{c^2}{9}\frac{(1 - p^4)^2}{p^2}$$

$$= \frac{c^2}{9}\frac{\left(1 - \frac{y^2}{x^2}\right)^2}{\left(-\frac{y}{x}\right)}$$

$$\Rightarrow \boxed{-y^3 = \frac{c^2 x}{9}\left(1 - \frac{y^2}{x^2}\right)^2}$$

WORKED EXAMPLE 41

Find the equation of the tangent to the rectangular hyperbola $xy = c^2$ at the point $P\left(cp, \frac{c}{p}\right)$. Find the area of the triangle formed by this tangent and the axes of coordinates. Find also the coordinates of the centroid of this triangle and hence its locus.

Solution 41

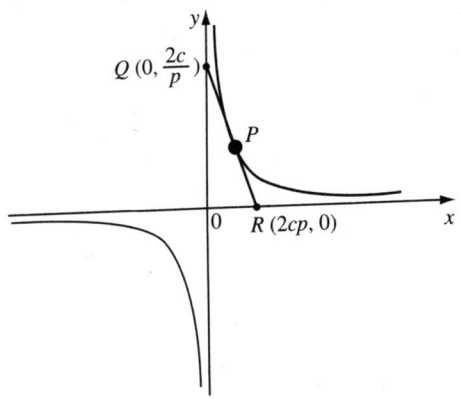

Fig. 6-I/98

The equation of the tangent is $x + p^2 y - 2cp = 0$.
This intersects the x and y-axis at R and Q repectively;
$R(2cp, 0)$ and $Q\left(0, \frac{2c}{p}\right)$.

Area of $\triangle OQR = \frac{1}{2} 2cp \frac{2c}{p} = 2c^2$.

The coordinates of the centre of gravity are $\left(\frac{2}{3} cp, \frac{2}{3} \frac{c}{p}\right)$.

The locus is found by eliminating p from $x = \frac{2}{3} cp$, $y = \frac{2}{3} \frac{c}{p}$ $\boxed{xy = \frac{4}{9} c^2}$ which is another rectangular hyperbola.

Central Rectangle of the Hyperbola

$$\frac{x^2}{a^2} - \frac{y^2}{b^2} = 1$$

$$y = \frac{b}{a} x$$

Asymptotes $y = -\frac{b}{a} x$

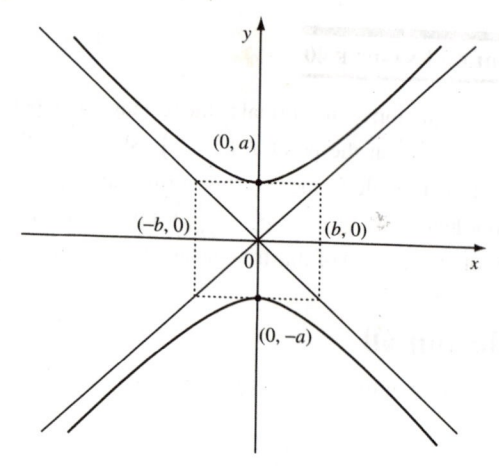

Fig. 6-I/100

WORKED EXAMPLE 42

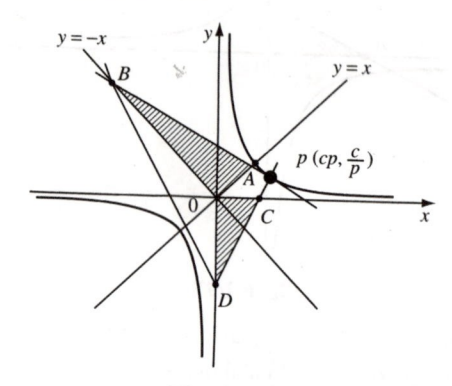

Fig. 6-I/101

The tangent at $P\left(cp, \frac{c}{p}\right)$ to the rectangular hyperbola $xy = c^2$ meets the lines $y = x$ and $y = -x$ at A and B respectively. Determine the area of the triangle $OAB(\Delta_1)$ where O is the origin. The normal at P intersects the x-axis and y-axis in C and D respectively. Determine the areas of the triangles $OCD(\Delta_2)$ and $BPD(\Delta_3)$. Find $\Delta_1 \Delta_2 \Delta_3$.

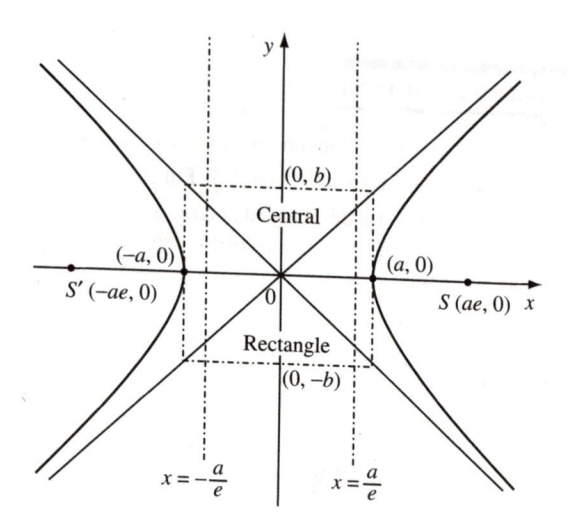

Fig. 6-I/99 The asymptotes of a hyperbola are the lines containing the diagonals of the central rectangle

$$\frac{y^2}{a^2} - \frac{x^2}{b^2} = 1$$

Asymptotes $y = \frac{a}{b} x$

$$y = -\frac{a}{b} x$$

Solution 42

$$x = cp, \quad \frac{dx}{dp} = c$$

$$y = \frac{c}{p} = cp^{-1}, \quad \frac{dy}{dp} = -cp^{-2} = -\frac{c}{p^2}$$

$$\frac{\frac{dy}{dp}}{\frac{dx}{dp}} = \frac{-\frac{c}{p^2}}{c} = -\frac{1}{p^2}.$$

The equation of the tangent at P is $y = mx + k$,

$y = -\dfrac{1}{p^2}x + k$,

$\dfrac{c}{p} = -\dfrac{1}{p^2}cp + k$,

$k = \dfrac{c}{p} + \dfrac{c}{p} = \dfrac{2c}{p}$

$y = -\dfrac{1}{p^2}x + \dfrac{2c}{p}$

$\boxed{yp^2 + x = 2cp}$

This tangent intersects $y = x$ and $y = -x$. The coordinates of A can be found by solving $yp^2 + x = 2cp$ and $y = x$.

$yp^2 + x = 2cp,\ y = \dfrac{2cp}{p^2 + 1} = x;$

$A\left(\dfrac{2cp}{p^2 + 1}, \dfrac{2cp}{p^2 + 1}\right).$

The coordinates of B can be found by solving $yp^2 + x = 2cp$ and $y = -x$.

$yp^2 + x = 2cp,$

$y = \dfrac{2cp}{p^2 - 1},$

$x = -\dfrac{2cp}{p^2 - 1}$

$B\left(-\dfrac{2cp}{p^2 - 1}, \dfrac{2cp}{p^2 - 1}\right).$

Area of triangle OAB

$= \dfrac{1}{2}\begin{vmatrix} 1 & 1 & 1 \\ 0 & \dfrac{2cp}{p^2-1} & \dfrac{2cp}{p^2+1} \\ 0 & \dfrac{2cp}{p^2-1} & \dfrac{2cp}{p^2+1} \end{vmatrix}$

$= \dfrac{1}{2}\begin{vmatrix} \dfrac{-2cp}{p^2-1} & \dfrac{2cp}{p^2+1} \\ \dfrac{2cp}{p^2-1} & \dfrac{2cp}{p^2+1} \end{vmatrix}$

$\Delta_1 = \dfrac{1}{2}\left(\dfrac{-2cp(2cp)}{(p^2-1)(p^2+1)} - \dfrac{2cp}{p^2-1}\cdot\dfrac{2cp}{p^2+1}\right)$

$\Delta_1 = \left(\dfrac{2c^2p^2}{(p^2-1)(p^2+1)} + \dfrac{2c^2p^2}{(p^2-1)(p^2+1)}\right)$

the area is $\Delta_1 = \dfrac{4c^2p^2}{p^4 - 1}$.

The equation of the normal at P

$y = p^2x + k$

$\dfrac{c}{p} = p^2 cp + k$

$k = \dfrac{c}{p} - cp^3$

$y = p^2x + \dfrac{c}{p} - cp^3$

$\boxed{py = p^3x + c - cp^4}$

when $y = 0$, the normal intersect at C,

$x = \dfrac{cp^4 - c}{p^3}$

$= cp - \dfrac{c}{p^3},\ C\left(cp - \dfrac{c}{p^3}, 0\right)$

when $x = 0$ the normal intersects at D,

$y = \dfrac{c}{p} - cp^3,\ D\left(0, \dfrac{c}{p} - cp^3\right)$

Area of triangle $OCD = \dfrac{1}{2}\begin{vmatrix} 1 & 1 & 1 \\ 0 & cp - \dfrac{c}{p^3} & 0 \\ 0 & 0 & \dfrac{c}{p} - cp^3 \end{vmatrix}$

$= \dfrac{1}{2}\left(cp - \dfrac{c}{p^3}\right)\left(\dfrac{c}{p} - cp^3\right)$

$= \dfrac{1}{2}c^2\left(\dfrac{p^4-1}{p^3}\right)\left(\dfrac{1-p^4}{p}\right)$

$\Delta_2 = \dfrac{1}{2}\dfrac{c^2(p^4-1)^2}{p^4}$

Area of BPD

$$= \frac{1}{2} \begin{vmatrix} 1 & 1 & 1 \\ \frac{-2cp}{p^2-1} & cp & 0 \\ \frac{2cp}{p^2-1} & \frac{c}{p} & \frac{c}{p}-cp^3 \end{vmatrix}$$

$$= \frac{1}{2}cp\left(\frac{c}{p}-cp^3\right) - \frac{1}{2}\left(-\frac{2cp}{p^2-1}\left(\frac{c}{p}-cp^3\right)\right)$$

$$+ \frac{1}{2}\left(-\frac{2cp}{p^2-1} \times \frac{c}{p} - cp\left(\frac{2cp}{p^2-1}\right)\right)$$

$$= \frac{1}{2}\left[c^2p\left(\frac{1-p^4}{p}\right) + \frac{2c^2p}{p(p^2-1)}\right.$$

$$\left. - \frac{2c^2p^5}{(p^2-1)p} - \frac{2c^2p}{p(p^2-1)} - \frac{2c^2p^2}{p^2-1}\right]$$

$$= \frac{1}{2}c^2\left[(1-p^4) + \frac{2}{p^2-1} - \frac{2p^4}{p^2-1}\right.$$

$$\left. - \frac{2}{p^2-1} - \frac{2p^2}{p^2-1}\right]$$

$$= \frac{1}{2}c^2\left(\frac{(1-p^4)(p^2-1) + 2 - 2p^4 - 2 - 2p^2}{p^2-1}\right)$$

$$= \frac{1}{2}\frac{c^2}{(p^2-1)}(p^2-1-p^6+p^4-2p^4-2p^2)$$

$$= \frac{1}{2}\frac{c^2}{(p^2-1)}(p^6+p^4+p^2+1)$$

$$\Delta_3 = \frac{1}{2}\frac{c^2}{(p^2-1)}(p^6+p^4+p^2+1)$$

$$\Delta_1\Delta_2\Delta_3 = \frac{4c^2p^2}{p^4-1} \cdot \frac{1}{2}\frac{c^2(p^4-1)^2}{p^4}$$

$$\cdot \frac{1}{2}\frac{c^2}{p^2-1}(p^6+p^4+p^2+1)$$

$$= \frac{c^6}{p^2}(p^2+1)(p^6+p^4+p^2+1).$$

Exercises 5

1. Sketch the following hyperbolas:
   (i) $\dfrac{x^2}{2^2} - \dfrac{y^2}{1^2} = 1$
   (ii) $\dfrac{x^2}{3^2} - \dfrac{y^2}{2^2} = 1$
   (iii) $\dfrac{x^2}{3^2} - \dfrac{y^2}{3^2} = 1.$

2. Determine the coordinates of (a) foci (b) vertices (c) the directrices and (d) eccentricities of the hyperbolas in question 1.

3. Sketch the rectangular hyperbolas:
   (i) $x^2 - y^2 = 1$
   (ii) $x^2 - y^2 = 9$.

   State the equations of the directrices and the eccentricities.

4. Verify that the parametric equations $x = a\sec t$, $y = b\tan t$ are those of the hyperbola $\dfrac{x^2}{a^2} - \dfrac{y^2}{b^2} = 1$ and state the coordinates of the foci and the equations of the asymptotes. If $a = b$ what are the new asymptotes?

5. The rectangular hyperbola has cartesian equation $xy = c^2$, write down the equations of the asymptotes, and check that the parametric equations are $x = ct$, $y = \dfrac{c}{t}$.

6. Show that the tangent at (x_1, y_1) of the general hyperbola $\dfrac{x^2}{a^2} - \dfrac{y^2}{b^2} = 1$ is given by $\dfrac{xx_1}{a^2} - \dfrac{yy_1}{b^2} = 1$.

7. Show that the tangent at $(a\sec\theta, b\tan\theta)$ of the hyperbola $b^2x^2 - a^2y^2 = a^2b^2$ is given by $bx\sec\theta - ay\tan\theta = ab$.

8. Show that the tangent at $(a\cosh u, b\sinh u)$ of the hyperbola $b^2x^2 - a^2y^2 = a^2b^2$ is given by $bx\cosh y - ay\sinh u = ab$.

9. Show that the foci of the rectangular hyperbola are given as $\left(\pm\sqrt{2}c, \pm\sqrt{2}c\right)$ and the equation of the directrices are $x + y = \pm c\sqrt{2}$.

10. Show that the tangent at (x_1, y_1) is $xy_1 + yx_1 = 2c^2$ of the rectangular hyperbola $xy = c^2$.

11. Show that the tangent at $\left(ct, \dfrac{c}{t}\right)$ is $x + t^2y = 2ct$ of the rectangular hyperbola $xy = c^2$.

12. Prove that the equation of the chord joining the points $R\left(cr, \frac{c}{r}\right)$ and $S\left(cs, \frac{c}{s}\right)$ on the rectangular hyperbola $xy = c^2$ is $rsy + x = c(r + s)$.

    It is given that RS subtends a right angle at the point $P\left(cp, \frac{c}{p}\right)$ on the curve. Show that RS is parallel to the normal at P to the curve.

13. Find the coordinates of the foci, the eccentricity and the length of the latus rectum of the hyperbola.
    $$\frac{x^2}{16} - \frac{y^2}{9} = 1.$$

14. Find the equations of the tangents to the hyperbola $x^2 - 4y^2 = 4$ which are perpendicular to the line $4x - 5y + 3 = 0$.

15. Show that the diameters $y = m_1 x$, $y = m_2 x$ of the hyperbola $b^2 x^2 - a^2 y^2 = a^2 b^2$ are conjugate if
    $$\boxed{m_1 m_2 = \frac{b^2}{a^2}.}$$
    (Two diameters of a hyperbola are said to be conjugate when each bisects all chords parallel to the other).

16. Show that the equation to the chord joining two points (x_1, y_1), (x_2, y_2) on the rectangular hyperbola $xy = c^2$ is $x(y_1 + y_2) + y(x_1 + x_2) = (x_1 + x_2)(y_1 + y_2)$. Hence show that as $x_2 \to x_1$ $y_2 \to y_1$ then the equation of the chord approaches that of the tangent $xx_1 + yy_1 = c^2$ at (x_1, y_1).

17. The centre of a hyperbola is the origin and its transverse axis lies along the x-axis. If the distance between the foci is $18\sqrt{3}$ and the eccentricity $\sqrt{3}$, write down the equation of the hyperbola.

18. The normal at $Q\left(q, \frac{1}{q}\right)$ to $xy = 1$ meets the curve at R. Find the length of QR in terms of q.

19. The coordinates of the foci of the hyperbola $b^2 x^2 - a^2 y^2 = a^2 b^2$ are $S(5, 0)$ and $S'(-5, 0)$ and the distance between the vertices is 8 find the equation of the hyperbola.

20. The equation of a hyperbola is $25x^2 - 16y^2 = 400$ determine the coordinates of the foci.

21. The equation of a hyperbola is $36x^2 - 49y^2 = 1764$ determine the coordinates of the foci.

22. A curve has the parametric equations $x = 4t$, $y = \frac{4}{t}$. Determine the equations of the tangent and normal to the curve at the point $T\left(4t, \frac{4}{t}\right)$.

    From the fixed point $A(-16, 8)$, two tangents are drawn to the hyperbola with points of contact B and C. Find the length of the chord BC, the perimeter of the triangle, ABC, and its area. The normals at B and C intersect at D find the coordinates of D and the area of the triangle BCD.

Miscellaneous

1. Show that the tangent at the point $P(3\cos\theta, 2\sin\theta)$ to the ellipse $4x^2 + 9y^2 = 36$ has equation $2x\cos\theta + 3y\sin\theta = 6$. (4 marks)

   Find the eccentricity of the ellipse and show that the distance between the foci, which are situated at the points S_1 and S_2 is $2\sqrt{5}$. (3 marks)

   The tangent at P meets the tangents which are drawn from the ends of the major axis in the points A and B, as shown in Fig. 6-M/1.

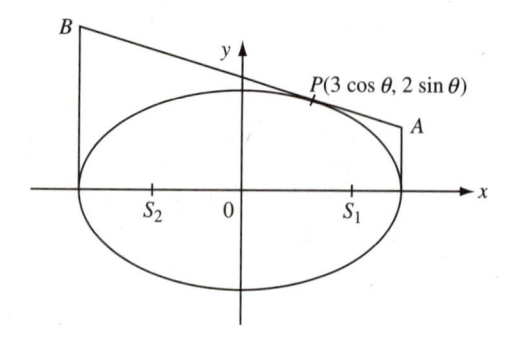

   Fig.6-M/1

   Show that the circle drawn on AB as diameter has area $\pi(5 + 4\csc^2\theta)$ and verify that the circle passes through S_1 and S_2. (8 marks)

2. Two points P and Q on the parabola $y^2 = 4ax$ have coordinates $(ap^2, 2ap)$ and $(aq^2, 2aq)$ respectively. Find

   (i) the slope of the chord PQ,

   (ii) the slope of the tangent at P to the parabola,

   (iii) the equation of the normal at P to the parabola. The normals at P and Q to the parabola intersect at R. Find the coordinates of R.

   If the chord PQ passes through the point $(-2a, 0)$, show that R lies on the parabola.

3. Find the equation of the tangent to the rectangular hyperbola $xy = c^2$ at the point $T\left(ct, \frac{c}{t}\right)$. If the tangent intersects the x-axis and y-axis at the points P and Q, show that T is the midpoint of PQ.

4. Show that the locus of $x = -ct$ and $y = \frac{c}{t}$ is given by the equation $xy + c^2 = 0$. Sketch this locus.

   The tangent at the point $P\left(-cp, \frac{c}{p}\right)$ cuts the x and y axes at the points A and B, find their coordinates. Determine the area of the triangle APB.

5. Find the equation of the circle C which has the line segment joining the points $(2, 2)$ and $(4, 6)$ as diameter. (3 marks)

   Calculate the distance of the centre of the circle C from the line $y + 2x = 5$ and hence show that this line is a tangent to the circle. (3 marks)

   Write down the equation of the line L of gradient m passing through the origin. Show that the x co-ordinates of the points of intersection of L and C satisfy the equation

   $(1 + m^2)x^2 - 2(3 + 4m)x + 20 = 0$ (4 marks)

   Write down the condition for the roots to be equal and hence find

   (a) the gradients of the two tangents from the origin to C,

   (b) the acute angle between the tangents.

   (6 marks)

   Ans. $(x - 3)^2 + (y - 4)^2 = 5$; $\sqrt{5}$; $y = mx$;

   (a) $\frac{1}{2}, 5\frac{1}{2}$; (b) $\tan^{-1}\frac{4}{3}$.

6. The point (x, y) moves in such a way that its distance from the point $(0, 1)$ is equal to its distance from the line $y = -1$. Show that the equation of the parabola generated is $x^2 = 4y$ (3 marks)

   Show that the equation of the normal to this parabola at the point $P(2p, p^2)$ is $x + yp = 2p + p^3$.

   (3 marks)

The line through P and the point $Q(0, -2)$ meets the parabola again at R. Show that

(a) the co-ordinates of R are $\left(\frac{4}{p}, \frac{4}{p^2}\right)$ (5 marks)

(b) the normals at P and R intersect at a point on the parabola. (4 marks)

7. Sketch the curve defined parametrically by
$x = 2 + t^2, y = 4t$ (2 marks)

Write down the equation of the straight line with gradient m passing through the point $(1, 0)$. (1 mark)

Show that this line meets the curve when
$mt^2 - 4t + m = 0$ (2 marks)

Find the values of m for which this quadratic equation has equal roots. Hence determine the equations of the tangents to the curve which pass through the point $(1, 0)$.

Ans. $m = \pm 2$, $y - 2x + 2 = 0$, $y + 2x - 2 = 0$
(3 marks)

8. The point P, Q and R have coordinates $(2, 4)$, $(8, -2)$ and $(6, 2)$ respectively.

(a) Find the equation of the straight line l which is perpendicular to the line PQ and which passes through the midpoint of PR. (3 marks)

(b) The line l cuts PQ at S. Find the ratio $PS: SQ$. (3 marks)

(c) The circle passing through P, Q and R has centre C. Find the coordinates of C and the radius of the circle. (5 marks)

(d) Given that angle $PCQ = \theta$ radians, show that $\tan \theta = \frac{24}{7}$.

Prove that the smaller segment of the circle cut off by the chord PQ has area $25\theta - 24$. (5 marks)

Ans. $y = x - 1$, $PS: SQ = 1: 3$, $C(1, -3)$, $5\sqrt{2}$

9. Given that a is a positive constant, sketch the curve with equation $ay = x^2$. (1 mark)

Show that the normal to the curve at the point $T(at, at^2)$ has equation $x + 2ty = at + 2at^3$. (3 marks)

Deduce that the normals to the curve at $A(a, a)$, $B(2a, 4a)$ and $C(-3a, 9a)$ all pass through a point D.

Find the coordinates of D. (4 marks)

Show that the tangent to the curve at T meets the y-axis at the point $V(0, -at^2)$. The finite region R is bounded by the part of the curve from the origin O to T, the line OV and the line VT. The region R is rotated completely about the y-axis. Find, in terms of π, t and a, the volume of the solid generated. (6 marks)

Ans. $D\left(-12a, \frac{15}{2}a\right)$, $V = \frac{1}{6}\pi a^3 t^4$

10. Show that an equation of the normal to the parabola with equation $y^2 = 4ax$, at the point $P(ap^2, 2ap)$ is $y + px = 2ap + ap^3$.

The normals at the points $P(ap^2, 2ap)$ and $Q(aq^2, 2aq)$ meet at the point N. Find the coordinates of N in terms of p, q and a. The points P and Q vary in such a way that $pq = -1$. Find a cartesian equation of the locus of N. Given that $p = 2$, find the length of the arc PQ of the parabola, giving your answer to 2 decimal places.

11.

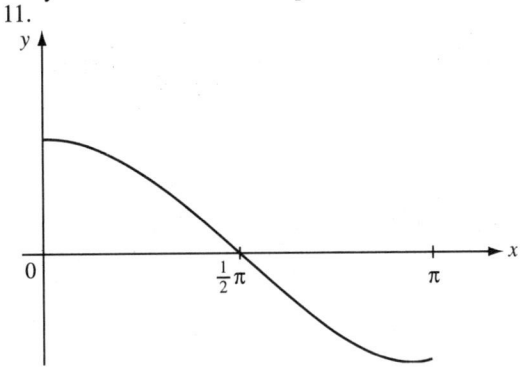

Fig.6-M/2

Figure 6-M/2 shows a sketch of the curve with equation

$$y = \frac{\cos x}{5 - \sin x}, 0 \leq x \leq \pi.$$

(a) By considering the values of $\sin x$ and $\cos x$ for which $\frac{dy}{dx} = 0$, show that

$$-\frac{1}{12}\sqrt{6} \leq y \leq \frac{1}{12}\sqrt{6}.$$

(b) Show that the normal to the curve at the point where $x = \frac{1}{2}\pi$ meets the y-axis at the point $(0, -2\pi)$.

The finite region R is bounded by the curve, the y-axis and this normal. (15 marks)

(c) Determine the area of R. (15 marks)

12. (a) Find the eccentricity of the ellipse with equation $3x^2 + 4y^2 = 12$.
    (b) Find an equation of the tangent to the ellipse with equation $3x^2 + 4y^2 = 12$ at the point with coordinates $\left(1, \frac{3}{2}\right)$. This tangent meets the y-axis at the point G.
    Given that S and S' are foci of the ellipse,
    (c) find the area of $\triangle SS'G$.
    (d) Show that the area of the surface of revolution generated when the ellipse with equation $3x^2 + 4y^2 = 12$ is rotated completely about the x-axis is given by
    $$\frac{\pi\sqrt{3}}{2} \int_{-2}^{2} \sqrt{16 - x^2}\, dx.$$
    Hence, evaluate this area.

13. Show that an equation of the tangent to the ellipse $\frac{x^2}{9} + \frac{y^2}{4} = 1$ at the point $P(3\cos\theta, 2\sin\theta)$ is $2x\cos\theta + 3y\sin\theta = 6$.
    Find, also, an equation of the normal to the ellipse at P. The tangent and normal at P meet the x-axis at T and N respectively. Find the coordinates of the centre C of the circle through the points P, T and N. Given that $OC:ON = 3:2$, where O is the origin, and that P is in the first quadrant, show that the coordinates of P are $\left(\frac{9}{\sqrt{10}}, \frac{2}{\sqrt{10}}\right)$.

14. With respect to a fixed origin O, the points A and B have coordinates $(-4, 0)$ and $(12, 12)$ respectively. The mid-point of AB is M. Find an equation of the line in the plane of the coordinate axes Ox and Oy which passes through M and is perpendicular to AB.
    Hence, or otherwise, find, in cartesian form, an equation of the circle which passes through O, A and B.
    (7 marks)

15. The parametric equations of a curve are $x = \cos t$, $y = 2\sin t$. Show that an equation of the normal to the curve at the point $P(\cos t, 2\sin t)$ is
    $$\frac{2y}{\sin t} - \frac{x}{\cos t} = 3.$$
    The normal to the curve at P meets the x-axis at G. The mid-point of PG is M. As t varies, find a cartesian equation of the locus of M. (8 marks)

16. Show that the hyperbola $x^2 - y^2 = a^2$, $a > 0$, has eccentricity equal to $\sqrt{2}$. Hence state the coordinates of the focus S and the equation of the corresponding directrix L, where S and L both lie in the region $x > 0$.

    The perpendicular from S to the line $y = x$ meets the line $y = x$ at P and the perpendicular from S to the line $y = -x$ meets the line $y = -x$ at Q. Show that both P and Q lie on the directrix L and give the coordinates of P and Q. Given that the line SP meets the hyperbola at the point R, prove that the tangent at R passes through the point Q.

17. A hyperbola has equation $x^2 - y^2 = 4$. Show that the point P with coordinates $x = t + \frac{1}{t}$, $y = t - \frac{1}{t}$ lies on the hyperbola (2 marks)

    Prove that the equation of the normal to the curve at P has equation $t(t^2+1)y + t(t^2-1)x = 2(t^4-1)$.
    (5 marks)

    The normal at P intersects the x-axis at X and the y-axis at Y.

    (a) Prove that the mid point of XY is P. (2 marks)

    (b) The mid-point of PX is M. Prove that the locus of M is a hyperbola and determine its eccentricity. (6 marks)

    Ans. (b) $\dfrac{x^2}{9} - y^2 = 1;\ \dfrac{1}{3}\sqrt{10}$

18. A curve is defined parametrically by
    $$x = \frac{2t}{1+t},\ y = \frac{t^2}{1+t}.$$
    Prove that the normal to the curve at the point $\left(1, \frac{1}{2}\right)$ has equation
    $$6y + 4x = 7.$$ (5 marks)
    Determine the coordinates of the other point of intersection of this normal with the curve (4 marks)

    Ans. $\left(14, -\dfrac{49}{6}\right)$

19. Given that t is a non-zero parameter, show that the point $P\left[2\left(t + \frac{1}{t}\right), \left(t - \frac{1}{t}\right)\right]$ always lies on the hyperbola $x^2 - 4y^2 = 16$. (2 marks)

    Show that the tangent at p to the hyperbola has equation $x(t^2+1) - 2y(t^2-1) = 8t$ (4 marks)

    The tangent at P meets the asymptotes of the hyperbola whose equations are $2y - x = 0$ and $2y + x = 0$ in the point L and M respectively. Find the coordinates of L and M in terms of t. (4 marks)

Show that, as t varies, the centre of the circle passing through L, M and the origin O lies on the hyperbola $4x^2 - y^2 = 25$. (5 marks)

Ans. $L(4y, 2t)$, $M\left(\dfrac{4}{t}, -\dfrac{2}{t}\right)$

20. The triangular region R has vertices A, B and C and is defined by $x \geq 0$, $y + 2 \geq 0$, $2y + x - 2 \leq 0$.

    (a) By drawing suitable straight lines on a sketch, show and label the region R.

    (b) Find the coordinates of A, B and C. The point P lies in R and is such that $PA = PB = PC$.

    (c) Determine the coordinates of P.

    (d) Hence find the radius of the circle with centre P which passes through A, B and C. (10 marks)

21. The curve C has parametric equations

    $x = t^3$, $y = 3t^2$, $t \geq 0$.

    (a) Find $\dfrac{dy}{dx}$ in terms of t.

    The points P and Q on C have parameters $t = 2$ and $t = 3$ respectively and O is the origin.

    (b) Show that the tangent to C at P is parallel to the line OQ. (7 marks)

22. Show that an equation of the normal to the parabola $y^2 = 4ax$ at the point

    $P(at^2, 2at)$ is $y + tx = 2at + at^3$.

    This normal meets the parabola again at the point $Q(as^2, 2as)$.

    Show that

    (a) $s + t + \dfrac{2}{t} = 0$, $t \neq 0$,

    (b) $PQ^2 = \dfrac{16a^2(t^2 + 1)^3}{t^4}$, $t \neq 0$. (17 marks)

23. The curve C has parametric equations

    $x = at$, $y = \dfrac{a}{t}$, $t \in R$, $t \neq 0$.

    where t is a parameter and a is a positive constant

    (a) Sketch C.

    (b) Find $\dfrac{dy}{dx}$ in terms of t.

    The point P on C has parameter $t = 2$.

    (c) Show that an equation of the normal to C at P is $2y = 8x - 15a$. This normal meets C again at the point Q.

    (d) Find the value of t at Q. (15 marks)

24. $x = 4c \cos t$, $y = 3c \sin t$, $-\pi < t < \pi$,

    where c is a positive constant.

    (a) Write down the value of t at the point $A(4c, 0)$ and at the point $B(0, 3c)$.

    (b) By considering the integral $\int y \dfrac{dx}{dt} dt$ find, in terms of c, the area of the region enclosed by the ellipse. (10 marks)

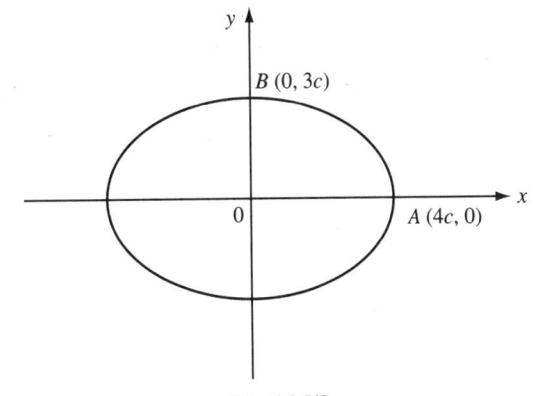

Fig.6-M/3

Fig. 6-M/3 shows a sketch of the ellipse given parametrically by the equations

25.

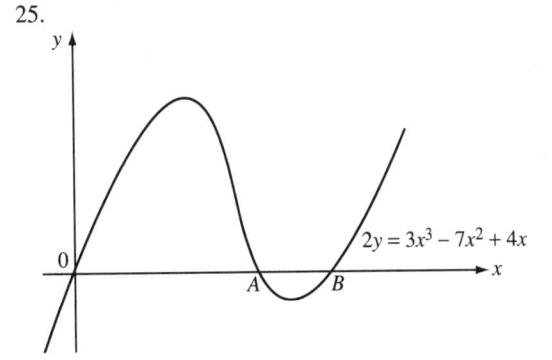

Fig.6-M/4

Figure 6-M/4 shows a sketch of part of the curve C with equation $2y = 3x^3 - 7x^2 = 4x$

Which means the x-axis at the origin 0, the point $A(1, 0)$ and the point B.

(a) Find the coordinates of B.

The normals to the curve C at the points O and A meet at the point N.

(b) Find the coordinates of N.

(c) Calculate the area of $\triangle OAN$.

At the points P and Q, on the curve C, $\frac{dy}{dx} = 0$.

Given that the x-coordinates of P and Q are x_1 and x_2,

(d) Find the values of

(i) $x_1 + x_2$,

(ii) $x_1 x_2$. (17 marks)

26. The points A and B have coordinates $(-4, 6)$ and $(2, 8)$ respectively. A line p is drawn through B perpendicular to AB to meet the y-axis at the point C.

(a) Find an equation of the line p.

(b) Determine the coordinates of C.

(c) Show, by calculation, that $AB = BC$. Given that $ABCD$ is a square whose diagonals intersect at the point M, calculate the coordinates of

(d) the point M,

(e) the point D. (15 marks)

Ans (a) $y + 3x - 14 = 0$,

(b) $(0, 14)$,

(d) $(-2, 10)$,

(e) $(-6, 12)$

27. An ellipse has parametric equations

$x = a \cos\theta$, $y = b \sin\theta$, $0 \leq \theta \leq 2\pi$, $a > b$,

where $b^2 = a^2(1 - e^2)$.

(a) Show that the length, s, of the arc of the ellipse from $\theta = 0$ to

$\theta = \pi$ is given by

$$s = a \int_0^\pi \sqrt{(1 - e^2 \cos^2\theta)} d\theta.$$

(b) Write down the first two non-zero terms in the expansion of $\sqrt{(1 - e^2 \cos^2\theta)}$, $0 < e < 1$, as a power series in ascending power of e. Given that e is sufficiently small so that powers of e higher than e^2 can be neglected,

(c) show that an approximate value of s is

$$\pi a \left(1 - \frac{1}{4}e^2\right). \qquad \text{(15 marks)}$$

28. Show that an equation of the tangent to the parabola with equation $y^2 = 4ax$ at the point $P(ap^2, 2ap)$ is $x - py + ap^2 = 0$.

The point $Q(aq^2, 2aq)$ on the parabola is such that OP is perpendicular to OQ. The tangents to the parabola at P and Q meet at the point R. Prove that, as P varies, the locus of R is a straight line, stating its equation. (14 marks)

29. The gradient of the curve C at a point (x, y) on the curve is given by the equation

$$\frac{dy}{dx} = x^2 + 2x.$$

The curve passes through the point with coordinates $(1, 2)$.

(a) Find an equation of C.

(b) Show that an equation of the normal n to C at the point $(1, 2)$ is $3y + x = 7$.

The line l is normal to C at the point (X, Y), $X \neq 1$, and also is parallel to n.

(c) Find the value of X and hence show that an equation for l is $3y + x + 1 = 0$.

The lines n and l cut the y-axis at the points A and B respectively. The point M lies on AB and is such that $AM : MB = 5 : 3$.

(d) Find the coordinates of the point M.

(e) Show that M lies on C. (17 marks)

30. A curve is defined for $x \neq 1$ by the equation

$$y = \frac{x + 2}{(x - 1)^2}.$$

Find the equation of the tangent to the curve of the point where $x = 0$. (4 marks)

Find the coordinates of the point where this tangent crosses the curve again. (4 marks)

Ans. $y = 5x + 2$, $\left(\dfrac{8}{5}, \dfrac{10}{3}\right)$

31. Show that the normal to the parabola with equation $y^2 = 4ax$ at the point $P(ap^2, 2ap)$ has equation
$$y + px = 2ap + ap^3.$$ (3 marks)

This normal meets the parabola again at the point $Q(aq^2, 2aq)$.

Show that $q = -\left(p + \dfrac{2}{q}\right)$.

The tangents to the parabola at P and Q meet at R. Determine the coordinates of R, in terms of a and p. Hence, show that R lies on the curve with equation
$$xy^2 + 4a^3 + 2ay^2 = 0.$$

Write down the equations of the two asymptotes to this curve. (8 marks)

Ans. $\left[-a(2+p^2), -\dfrac{2a}{p}\right]$; $x = -2a$, $y = 0$

32. Determine the coordinates of the centre and the radius of the circle with equation
$$x^2 + y^2 + 2x - 6y - 26 = 0.$$ (3 marks)

Find the distance from the point $P(7, 9)$ to the centre of the circle. Hence find the length of the tangents from P to the circle. (4 marks)

Ans. $(-1, 3)$, $r = 6$; length are 10, 8.

33. Sketch the curve C defined parametrically by
$$x = t^2 - 2; \quad y = t.$$ (2 marks)

Write down the cartesian equation of the circle with centre the origin and radius r. Show that this circle meets the curve C at points whose parameter t satisfies the equation
$$t^4 - 3t^2 + 4 - r^2 = 0.$$ (2 marks)

(a) In the case $r = 2\sqrt{2}$, find the coordinates of the two points of intersection of the curve and the circle. (3 marks)

(b) Find the range of values of r for which the curve and the circle have exactly two points in common.

Ans. (a) $(2, 2)$, $(2, -2)$;

(b) $r > 2 \left(\text{also } r = \dfrac{\sqrt{7}}{2}\right)$.

34. Show that the tangent to the ellipse $\dfrac{x^2}{a^2} + \dfrac{y^2}{b^2} = 1$ at the point

$P(a \cos\theta, b \sin\theta)$ has equation
$$ay \sin\theta + bx \cos\theta = ab.$$ (4 marks)

The tangent at P meets the x-axis at A and the y-axis at B, and Q is the mid-point of AB.

Write down the coordinates of A, B and Q.

(a) Show that the distance PQ is given by
$$PQ^2 = \cot^2 2\theta (a^2 \sin^2\theta + b^2 \cos^2\theta).$$
(6 marks)

(b) Find, in terms of a and b, the coordinates of the four positions of the point P for which P itself is the mid-point of AB. (2 marks)

(c) Find a cartesian equation of the locus of Q as θ varies. (3 marks)

Ans (b) $\left(\pm\dfrac{a}{\sqrt{2}}, \pm\dfrac{b}{\sqrt{2}}\right)$;

(c) $b^2 x^2 + a^2 y^2 = 4x^2 y^2$.

35. Determine the coordinates of the centre C and the radius of the circle with equation
$$x^2 + y^2 + 4x - 6y = 12.$$ (2 marks)

The circle cuts the x-axis at the points A and B. Calculate the area of the triangle ABC. (4 marks)

Calculate the area of the minor segment of the circle cut off by the chord AB, giving your answer to three significant figures.

36. Find the equation of the chord joining the two points $P\left(cp, \dfrac{c}{p}\right)$ and $Q\left(cq, \dfrac{c}{q}\right)$ on the rectangular hyperbola $xy = c^2$.

By considering a suitable limit, deduce that the tangent to the curve at the point $T\left(ct, \dfrac{c}{t}\right)$ has equation $t^2 y + x = 2ct$.

(a) The point $R\left(ct, \dfrac{c}{r}\right)$ also lies on the hyperbola.

(i) Prove that P, Q and R cannot lie on the same straight line. (2 marks)

(ii) Show that, when angle $PQR = 90°$, $pq^2 r = -1$ and deduce that the tangent to the hyperbola at Q is perpendicular to the line PR.

(b) The tangent at T cuts the y-axis at Y and M is the mid-point of TY. Determine a cartesian equation of the locus of M as t varies.

(5 marks)

37. Prove that the tangent to the rectangular hyperbola $xy = c^2$ at the point $P(cp, \frac{c}{p})$ has equation

$$x + yp^2 = 2cp.$$ (3 marks)

Find the equation of the normal at P and prove that this normal meets the curve again at $Q(-cp^3, -\frac{c}{p^3})$. (5 marks)

The tangents at P and Q intersect at R. Find the coordinates of R in terms of c and p. (4 marks)

Hence prove that as p varies the locus of R has equation

$$(x^2 - y^2)^2 + 4c^2 xy = 0.$$ (3 marks)

Ans. $p^3 x - py = c(p^4 - 1)$;

$$x_R = -\frac{2cp}{(p^4 - 1)}, \quad y_R = \frac{2cp^3}{(p^4 - 1)}.$$

38. Find the equation of the tangent to the ellipse $\frac{x^2}{16} + y^2 = 1$ at the point $(4\cos t, \sin t)$. (2 marks)

A tangent to this ellipse cuts the coordinate axis in the points Q and R. Show that for all such tangents, the mid-point of QR lies on the curve

$$\frac{16}{x^2} + \frac{1}{y^2} = 4.$$ (4 marks)

Draw a sketch of this curve and state the equation of the four asymptotes (3 marks)

Show that an equation of this curve in polar coordinates form is

$$r^2 = 4\sec^2\theta + \frac{1}{4}\csc^2\theta.$$ (2 marks)

Find the least distance of the curve from the origin O.

(4 marks)

Ans. $4y\sin t + x\cos t = 4$, $x = \pm 2$, $y = \pm\frac{1}{2}$, 2.5

39. An ellipse has equation $16x^2 + 25y^2 = 400$. Determine its eccentricity and show that the equation of the normal to the ellipse at the point

$P(5\cos\theta, 4\sin\theta)$ is

$4y\cos\theta - 5x\sin\theta + 9\sin\theta\cos\theta = 0$. (6 marks)

(a) The normal at P cuts the coordinate axes at the points Q and R. The midpoint of QR is M. Determine the coordinates of M and hence deduce that the locus of M is an ellipse with the same eccentricity as the original ellipse.

(5 marks)

(b) If the normal at P passes throught the point $(5, 4)$, show that θ satisfies the equation $32\cos\theta - 50\sin\theta + 9\sin 2\theta = 0$. Using $\frac{\pi}{4}$ as a first approximation to a roof of this equation, obtain a second estimate by the Newton-Raphson method, giving your answer to two decimal places. (4 marks)

Ans. $e = 0.6$ (a) $M\left(\frac{9}{10}\cos\theta - \frac{9}{8}\sin\theta\right)$

(b) 0.72

40. The circle with equation $(x-5)^2 + (y-7)^2 = 25$ has centre C. The point $P(2, 3)$ lies on the circle. Determine the gradient of PC and hence, or otherwise, obtain the equation of the tangent to the circle at P. (4 marks)

Find also the equation of the straight line which passes through the point C and the point $Q(-1, 4)$. The tangent and line CQ intersect at R. Determine the size of angle PRC, to the nearest $0.1°$.

(4 marks)

Ans. $y - 3 = -\frac{3}{4}(x - 2)$; $63.4°$.

41. Prove that the equation of the normal to the parabola $y^2 = 4ax$ at the point $P(ap^2, 2ap)$ is

$y + px = 2ap + ap^3$. (3 marks)

(a) Find the point of intersection R of the normal at P and the normal at $Q(aq^2, 2aq)$.

(4 marks)

Given that the straight line through P and Q passes through the point $(-2a, 0)$ show that $pq = 2$ and hence deduce that R lies on the parabola. (4 marks)

(b) Let S be the point $(as^2, 2as)$ on the parabola If $s^2 > 8$, show that there are three distinct normals to the parabola which pass through S.

Ans. (a) $[a(2 + p^2 + q^2 + pq),$

$-apq(p+q)]$

42. Given that c is positive constant, sketch the ellipse whose equation is $16x^2 + 25y^2 = 30c^2$. Mark on your sketch the coordinates of the points where the ellipse crosses the coordinate axes Ox and Oy. Show that the foci of the ellipse are situated at $S(3c, 0)$ and $F(-3c, 0)$. (4 marks)

The tangent to the ellipse at $P\left(3c, \frac{16c}{5}\right)$ meets the x-axis at A and the y-axis at B. The normal to the ellipse at P meets the x-axis at C and the y-axis at D. Determine, in terms of c, the coordinates of $A, B, C,$ and D. (4 marks)

Prove that $\dfrac{\text{Area of triangle } OAB}{\text{Area of triangle } OCD} = \dfrac{5^6}{3^6}$. (3 marks)

The mid-point of FP is M. Show that M is the centre of the circle passing throught the points D, F, P and S. (4 marks)

Ans. $A\left(\dfrac{25c}{3}, 0\right)$, $B(0, 5c)$, $C\left(\dfrac{27c}{25}, 0\right)$,

$D\left(0, -\dfrac{9c}{5}\right)$

43. The straight lines $3x + 4y = 14$ and $x - 7y = 13$ intersect at the point A, and the line $4x - 3y + 23 = 0$ cuts the other two lines at the points B and C respectively.

   (a) Show that the length of BC is 10 units (4 marks)

   (b) Calculate the value of the acute angle BAC (3 marks)

   (c) Find the coordinates of A and calculate the perpendicular distance from A to BC. Hence find the equation of the circle with centre A for which BC is a tangent. (5 marks)

   (d) Find, in terms of p, the area of that part of the triangle ABC which lies outside the circle. (4 marks)

Ans. (b) 45°

(c) $(6, -1)$, 10, $(x-6)^2 + (y+1)^2 = 100$

(d) $50 - \dfrac{25\pi}{2}$.

44. Referred to an origin O and coordinate axes Ox and Oy, a curve is given by

$$x = \sec t + \tan t, \quad y = \csc t + \cot t, \quad 0 < t < \dfrac{\pi}{2},$$

where t is a parameter.

Prove that $\dfrac{dy}{dx} = \dfrac{1 - \sin t}{1 - \cos t}$. (5 marks)

Show that the normal to the curve at the point S, where $t = \tan^{-1}\left(\dfrac{3}{4}\right)$, has equation $x - 2y + 4 = 0$. Find the equation of the normal to the curve as the point T, where $t = \tan^{-1}\left(\dfrac{4}{3}\right)$. These normals meets at the point N.

Find the coordinates of N. (5 marks)

Hence calculate

   (a) the area of triangle SNT, (3 marks)

   (b) the tangent of the angle SNT. (3 marks)

Ans. $(4, 4)$ (a) 1.5 (b) 0.75

6. Answers

Exercises 1

65. $p = \pm \dfrac{ax_1 + by_1 + c}{\sqrt{a^2 + b^2}}$

66. $y = x, y = -x$

67. $x = 0, y = 0$

68. $14x - 112y - 21 = 0, 64x + 8y - 31 = 0$

69. $\dfrac{a_1 x + b_1 y + c_1}{\sqrt{a_1^2 + b_1^2}} = \pm \dfrac{a_2 x + b_2 y + c_2}{\sqrt{a_2^2 + b_2^2}}$

70. $a_1 x + b_1 y + c_1 + k(a_2 x + b_2 y + c_2) = 0$

71. $x(1 - 2k) + y(1 + k) + 3 + 6k = 0$

Exercises 2

1. $x^2 + y^2 = 4$

2. $x^2 + y^2 = 25$

3. $x^2 + y^2 = 49$

4. $c(0, 0), r = 2$

5. $x^2 + 2gx + y^2 + 2yf + c = 0$

6. $19x^2 - 21x + 19y^2 - 103y = 0$

7. $x^2 - 1.78x + y^2 + 2.32y - 3.36 = 0$

8. $x^2 + (y - 6)^2 = 4^2$

9. $x^2 - 6x + y^2 - 4y - 5 = 0$

10. (i) $x^2 + (y - 3)^2 = 3^2$
    (ii) $(x + 1)^2 + (y - 2)^2 = 2^2$
    (iii) $(x - 2)^2 + (y + 3)^2 = 1^2$

11. $\left(x + \dfrac{1}{2}\right)^2 + (y + 1)^2 = \dfrac{1}{2^2}$

12. $\left(x + \dfrac{5}{8}\right)^2 + \left(y + \dfrac{3}{8}\right)^2 = 2^2$

13. $(1, 3), (2.6, 2.2)$

14. $c(4, 4), r = \sqrt{10}$

18. $x^2 + y^2 + 6x + 8y - 1 = 0$

19. $x^2 + y^2 - 2\sqrt{5}x - 2\sqrt{3}y + 6 = 0$

20. $x^2 + 2gx + y^2 + 2fy + c = 0$

21. $(1 + m^2)(k^2 + 2kf + c) = (mk + g + mf)^2$

22. $3x^2 + 3y^2 + 11x + 9y - 44 = 0$

23. $2x^2 + 2y^2 - 13y = 0$

24. $x^2 + y^2 = 5^2$

25. $x^2 + y^2 + 6x - 8y + 16 = 0$

26. $x^2 + y^2 + 2gx + 2fy + c = 0$

27. $x^2 + y^2 - 4x - 6y + 12 = 0$

28. $xx_1 + yy_1 + g(x + x_1) + f(y + y_1) + c = 0$

31. (i) $x = 2$
    (ii) $y = -x - 2$

32. $9x^2 + 9y^2 - 74x - 58y = 0$

33. $x^2 + y^2 + x - y - 18 = 0$

34. $(x - 1)^2 + (9y + 3)^2 = 50, c(1, -3), r = \sqrt{50}$

35. $(y - y_1)(y - y_2) + (x - x_1)(x - x_2) = 0$

36. $C(2, 1), r = 3$

(i) 224 square units

(ii) 120 square units, 634 square units

Exercises 3

6. $y = tx - at^2$

7. $y = -px + 2ap + ap^3$

8. $S(a, 0), x = -a$

9. $S(-a, 0), x = a$

10. $S(0, a), y = -a$

11. $S(0, -a), y = a$

12. (i) $y^2 = -8x$
    (ii) $y^2 = 12x$
    (iii) $x^2 = 4y$
    (iv) $x^2 = -16y$

13. $c = -am^2, x = 2am, y = am^2$

14. $x^2 = -a(y + 3a)$

18. $2(x - aq^2) = (y - 2aq)(p + q)$

19. $T[-apq, a(p+q)]$

20. $y = 4.19x - 17.6a, y = -1.19x - 1.43a$

21. $S(0, 5), y = 3, AB = 4$

22. (i) $AB = 8$
    (ii) $AB = 4$
    (iii) $AB = 8$
    (iv) $AB = 4$

23. $y = 1$

25. $S\left(-1, \dfrac{9}{4}\right), y = \dfrac{7}{4}$

26. $x^2 = 12y$

27. $S\left(-\dfrac{3}{4}, -5\right), x = -\dfrac{5}{4}$

28. $T\left[2(p^2 + pq + q^2) + 4, -pq(2p+q)\right]$

29. (i) $y^2 = 8x$
    (ii) $x^2 = 12y$
    (iii) $(y-2)^2 = -8(x-3)$
    (iv) $x^2 = -20y$

(v) $(y-4)^2 = 8(x-4)$

(vi) $(x+3)^2 = -8(y+4)$

31. $x = 2a + a(p^2 + q^2 + pq), y = -apq(p+q)$

32. $x^2 = 12y$

Exercises 4

1. (ii) $a(\sin\theta)y + b(\cos\theta)x = ab$
   (iii) $b y \cos\theta = a(\sin\theta)x + (b^2 - a^2)\sin\theta\cos\theta$
   (iv) $y = -\dfrac{b}{a}(\cot\theta)x$
   (v) $y = \dfrac{a}{b}(\tan\theta)x$

2. $y = m_1 x + c, y = m_2 x + c$ where m_1, m_2 are given by
$$m = \dfrac{2x_1 y_1 \pm \sqrt{4x_1^2 y_1^2 - 4(x_1^2 - a^2)(y_1^2 - b^2)}}{2(x_1^2 - a^2)}$$
and the points of contacts

are $P\left(-\dfrac{a^2 m_1}{c}, \dfrac{c^2 - a^2 m_1^2}{c}\right)$ and

$Q\left(-\dfrac{a^2 m_2}{c}, c^2 - \dfrac{a^2 m_1^2}{c}\right)$

3. $T\left(3\sec\alpha - \dfrac{3\sin\frac{(\alpha+\beta)}{2}}{\cos\frac{(\alpha-\beta)}{2}}\tan\alpha, \dfrac{4\sin\frac{(\alpha+\beta)}{2}}{\cos\frac{(\alpha-\beta)}{2}}\right)$

5. $y = 8.11x - 40.7, y = 0.62x + 4.28$

6. (i) (a) $(\pm 1, 0)$
      (b) $e = \pm\dfrac{1}{2}$
      (c) $x = \pm 4$

   (ii) (a) $\left(\pm\dfrac{1}{\sqrt{2}}, 0\right)$
        (b) $e = \pm\dfrac{1}{\sqrt{2}}$
        (c) $x = \pm\sqrt{2}$

   (iii) (a) $\left(\pm\dfrac{3\sqrt{7}}{4}, 0\right)$

(b) $e = \pm\dfrac{\sqrt{7}}{4}$

(c) $x = \pm\dfrac{12}{\sqrt{7}}$

(iv) (a) $(\pm 4, 0)$

(b) $e = \pm\dfrac{4}{5}$

(c) $\pm\dfrac{25}{4}$

(v) (a) $(\pm 2, 0)$

(b) $e = \pm\dfrac{1}{\sqrt{2}}$

(c) ± 4

7. (i) (a) $(\pm 2, 0)$

(b) $e = \pm\dfrac{1}{\sqrt{3}}$

(c) $x = \pm 6$

(ii) (a) $\left(\pm\dfrac{4\sqrt{7}}{\sqrt{8}}, 0\right)$

(b) $e = \pm\dfrac{\sqrt{7}}{\sqrt{8}}$

(c) $x = \pm\dfrac{8\sqrt{2}}{\sqrt{7}}$

(iii) (a) $\left(\pm\dfrac{\sqrt{3}}{2}, 0\right)$

(b) $e = \pm\dfrac{\sqrt{3}}{2}$

(c) $x = \pm\dfrac{2}{\sqrt{3}}$

8. (i) $\dfrac{2x}{5} + \dfrac{\sqrt{5}y}{25} = 1$

(ii) $-\dfrac{2x}{8} + \dfrac{y}{2} = 1$

9. $4a^2x^2 + 4b^2y^2 = (a^2 - b^2)^2$

11. $y(a\cos\theta + ae) = b\sin\theta(x + ae)$

13. $x^2 + y^2$

$= \cos^2\dfrac{\phi - \theta}{2}\left(a^2\cos^2\dfrac{\phi+\theta}{2} + b^2\sin^2\dfrac{\phi+\theta}{2}\right)$

15. (i) $y = \dfrac{16}{25}x - \dfrac{82}{25}, y = -\dfrac{25}{16}x + \dfrac{9}{8}$

(ii) $y = 3x + 9, 3y + x = 7$

(iii) $y = -\dfrac{9}{8}x + 25, y = \dfrac{8}{9}x + \dfrac{20}{9}$

(iv) $y = x + 4, y = -x + 2$

(v) $4y + 3x = 9, 9y = 12x + 14$

16. $y = 0; (2, 0)$ 17. $A\left(\dfrac{9}{5}, \dfrac{16}{5}\right), B\left(-\dfrac{9}{5}, -\dfrac{16}{5}\right)$

18. $y = -3x - 13, y = -3x + 13$

20. $2y + x = 5, \quad 4y = 5x + 17, \quad Q(-2.65, 0.94),$
$P(1.8, 1.6)$

Exercises 5

2. (i) (a) $\left(\pm\dfrac{2\sqrt{5}}{2}, 0\right)$

(b) $(\pm 2, 0)$

(c) $\pm\dfrac{4}{\sqrt{5}}$

(d) $\pm\dfrac{\sqrt{5}}{2}$

(ii) (a) $\left(\pm\dfrac{3\sqrt{13}}{3}, 0\right)$

(b) $(\pm 3, 0)$

(c) $\pm\dfrac{9}{\sqrt{13}}$

(d) $\pm\dfrac{\sqrt{13}}{3}$

(iii) (a) $(\pm 3\sqrt{2}, 0)$

(b) $(\pm 3, 0)$

(c) $\pm\dfrac{3}{\sqrt{2}}$

(d) $\pm\sqrt{2}$

3. (i) $x = \pm\dfrac{1}{\sqrt{2}}, e = \pm\sqrt{2}$

(ii) $x = \pm\dfrac{3}{\sqrt{2}}, e = \pm\sqrt{2}$

4. $y = \pm x$

5. $x = 0, y = 0$

13. $S(\pm 5, 0), x = \pm \dfrac{16}{5}, AB = 4.5$

14. $4y + 5x = 7.12, 4y + 5x = -7.12$

17. $\dfrac{x^2}{9^2} - \dfrac{y^2}{\frac{81}{2}} = 1$

18. $RQ = \dfrac{(q^4 + 1)^{\frac{3}{2}}}{q^3}$

19. $\dfrac{x^2}{16} - \dfrac{y^2}{9} = 1$

20. $S(\sqrt{41}, 0), S'(-\sqrt{41}, 0)$

21. $S(\sqrt{85}, 0), S'(-\sqrt{85}, 0)$

22. $y = 4x - 30, y = x, 13.4, 55.1, 110, D(10, 10), 44$

6. COORDINATE GEOMETRY IN TWO DIMENSIONS
Index

Abscissae 1
Acute angle 5
Angle
    acute 5
    between curves 31
    between lines 5–7, 12
    eccentric 34
    obtuse 6–7
    tangent 6

Area
    ellipse 35
    triangle 9

Asymptotes
    to the hyperbola 41–2, 44

Auxiliary circle 34

Axes
    cartesian 1
    rectangular x-axis 1
    y-axis 1
    symmetry 34

Axis of conic
    ellipse 33
    hyperbola 40
    parabola 25–6

Centroid
    triangle 9

Central Rectangle of the hyperbola 56

Circle
    auxiliary 34
    centre, radius 18
    director 38
    equation 18
    general equation 18–20
    tangent, equation 19
    orthogonal 20
    parametric 22

Conic asymptote
    axes 41–4
    centre 18, 33, 39, 46
    conjugate diameters 58
    director circle 38
    directrix 25, 33, 39

Conic
    focus 25, 33, 39
    ellipse 33
    hyperbola 39
    parabola 25

Conjugate
    axis 46
    diameters 58–9
    hyperbolas 53

Coordinates
    mid-point 8, 12
    parametric 26, 30, 34, 36–7, 39–40

Curve
    circle 18
    parabola 25
    ellipse 33
    hyperbola 39

Descartes (Cartesian geometry) 1

Diameter
    circle 18
    conjugate
    ellipse 34
    hyperbola 53
    parabola 29
    rectangular hyperbola 52

Director circle 38

Directrix 25, 33, 39

Distance from a known point to a known line between two points 10–11

Division in a given ratio 7, 12
Double Ordinate 34

INDEX

Eccentric angle 34
Eliminating the parameter 29
Eccentricity 33, 39
Eccentricity of circle 33
Eccentricity of parabola 33
Eccentricity of ellipse 33
Eccentricity of hyperbola 39

Ellipse
    area 35
    auxiliary circle 34
    centre, diameter, vertices 34
    chord 36–7

Ellipse
    director circle 38
    directrix 33
    eccentric angle 34
    focus equation 33
    focal chord 33
    normal 36
    parametric equations 36
    principal axes 34
    tangent 35

Equation of locus
    circle 18–19
    ellipse 33–4
    hyperbola 39–40
    line 1–3
    parabola 25

Equations
    linear 1–5
    parametric 21, 26, 35
    perpendicular lines 12

Focal chord 28–9, 37

Focus
    ellipse 33
    hyperbola 39
    parabola 25

Gradient of a straight line 1
    General form of a line 4
    Gradient Intercept form 2

Heron's Formula 9

Hyperbola
    asymptotes 41–2
    centre, diameter chord 46
    conjugate axis 46
    conjugate diameters 46
    conjugate hyperbola 53
    directrix 39
    eccentricity 39
    focus 39
    general equation 39–40
    hyperbolic
    normal 40
    parametric equations 40
    trigonometric 39
    tangent 40
    vertices 39

Intercept form of a line 3

Latus rectum 58–9
Length of a distance 1, 11

Lines
    gradient 1–3, 11
    intercept form 3, 11
    angle between equation 5
    parallel 12
    perpendicular 12
    general form 1, 11
    gradient/intercept 2

Locus 25, 33, 39

Major axis 34
Medians
    intersection 9
Mid-point 8–9, 12

Normal to Hyperbola 48
Normals 27, 35, 48
Normal to the ellipse 36
Normal to the parabola 27–8

Obtuse 5
Ordinates 1
Ordinate (double) 34
Origin 1
Orthocentre 15
Orthogonal
    curves 20
    circles 20

Parabola
    axis 25
    chord 26
    contact of chord 26–7
    directrix 25
    equation 25
    focal chord 26, 28

focus 25
  normal 27
  properties 25
  section of cone 25
  tangent 27
  vertex 25
Parallel lines 12
Parametric equations 21, 26–32, 35, 39
Perpendical distance from a point to a line 10
Principal axes 34

Rectangular hyperbola
  asymptotes 44
  axes 41–2
  centre 41
  chord 45
  conjugate axis 46
  conjugate diameters 53
  conjugate hyperbolas 53
  diameter 52
  equation 46–56
  geometry of normal 48
  major axis 46
  parametric equations 47
  transverse axis 50
  tangent 50
  vertices 50

Tangent
  to conic 19
  to ellipse 35
  to hyperbola 40
  to parabola 26–8

Triangle
  area 8–9
  centroid, medians 9